Handbook of Frozen Food Processing and Packaging

Handbook of Frozen Food Processing and Packaging

Gordon Langowski

Oxford Book Company
Jaipur ▪ New Delhi

Published By

Oxford Book Company

Regd. Off. "Bony Residency", Gate No. 2
Opp. Tilak Public School, Vishweswariya Nagar,
Gopalpura Road, Jaipur - 302018, India
Ph: 0141 - 2761280, Telefax: 2761381
email - oxfordbookjpr@gmail.com
website - oxfordbookcompany.com

Delhi Office: 102, 1st Floor, Satyam House,
4327/3, Ansari Road, Darya Ganj,
New Delhi - 110002, India
Ph: 011-45652440

Distributed By

KEYA INTERNATIONAL PRESS LLP
E-mail: keyainternationalpressllp@gmail.com

Published in India, 2025

ISBN : 9789350303795

Typeset by Shivangi Computers, Jaipur

Printed and bound in India by Thomson Press Ltd.,
New Delhi

Contents

Preface

Consumer demand for a year-round supply of seasonal produce and ready-made meals remains the driving force behind innovation in frozen food technology. Freezing food preserves it from the time it is prepared to the time it is eaten. Since early times, farmers, fishermen, and trappers have preserved their game and produce in unheated buildings during the winter season. Freezing food slows down decomposition by turning residual moisture into ice, inhibiting the growth of most bacterial species. In the food commodity industry, there are two processes: mechanical and cryogenic (or flash freezing). The freezing kinetics is important to preserve the food quality and texture. Quicker freezing generates smaller ice crystals and maintains cellular structure. Cryogenic freezing is the quickest freezing technology available due to the ultra low liquid nitrogen temperature “196 °C (“320.8 °F)}. Preserving food in domestic kitchens during the 20th and 21st centuries is achieved using household freezers. Accepted advice to householders was to freeze food on the day of purchase. The freezing technique itself, just like the frozen food market, is developing to become faster, more efficient and more cost-effective.

Mechanical freezers were the first to be used in the food industry and are used in the vast majority of freezing / refrigerating lines. They function by circulating a refrigerant, normally ammonia, around the system, which withdraws heat from the food product. This heat is then transferred to a condenser and dissipated into air or water. The refrigerant itself, now a high pressure, hot liquid, is directed into an evaporator. As it passes through an expansion valve, it is cooled and then vaporises into a gaseous state. Now a low pressure, low temperature gas again, it can be reintroduced into the system. Cryogenic or (flash freezing) of food is a more recent development, but is used by many leading food manufacturers all over the world. Cryogenic equipment uses very low temperature gases – usually liquid nitrogen or solid carbon dioxide – which are applied directly to the food product. Frozen food packaging must maintain its integrity throughout machine filling, sealing, freezing, storage, transportation, thawing, and often cooking. As many frozen foods are cooked in a microwave oven, manufacturers have developed packaging that can go straight from freezer to the microwave. In 1974, the first differential heating container (DHC) was sold to the public. A DHC is a sleeve of metal designed to allow frozen foods to receive the correct amount of heat. Various sized apertures were positioned around the sleeve. The consumer would put the frozen dinner into the sleeve according to what needed the most heat. This ensured proper cooking. Today there are multiple options for packaging frozen foods. Boxes, cartons, bags, pouches, heat-in-bag pouches, lidded trays and pans, crystallized PET trays, and composite and plastic cans. Scientists are continually researching new aspects of frozen food packaging. Active packaging offers a host of new technologies that can actively sense and then neutralize the presence of bacteria or other harmful species. Active packaging can extend shelf-life, maintain product safety, and help preserve the food over a longer period of time. The speed of the freezing has a direct impact on the size and the number of ice crystals formed within a food product's cells and extracellular space. Slow freezing leads to fewer but larger ice crystals while fast freezing leads to smaller but more numerous ice crystals. Large ice crystals can puncture the walls of the cells of the food product which will cause a degradation of the texture of the product as well as the loss of its natural juices during thawing.

The book is designed to be a valuable source and reference book for scientists, product developers, students and all professionals with an interest in this field.

— **Gordon Langowski**

Chapter 1

Planning and Management of Food Services

In the past few decades, people ate in restaurants occassionally to celebrate a special event such as an anniversary, a birthday, or an achievement. It was an outing to look forward to, and if one indulged, it did not matter as these outings were rare. Today, the scenario is different. Eating out has become a way of life. Education and employment has taken many of us away from home, and the mother's role now has an added responsibility of contributing to the family income. Modern day compulsions have made eating out a necessity. No longer does one find time for the traditional fare of yesteryears and depends on the caterer for the following:

1. Food for festivals and celebrations
2. Meals at the work place.
3. Ready-to eat meals picked up on the way home from work
4. Snacks and sweetmeats for daily consumption
5. Preserves, pickles, papads, etc.
6. All meals served in institutions such as hospitals, school/college cafeteria, mess or dining hall, and boarding schools.

The number of reported cases of diabetes, hypertension, obesity, heart attacks, etc. is on the rise and so is the number of meals consumed away from home. This is not surprising because if one indulges practically everyday, it is bound to result in ill-health because of malnutrition. The caterer's role has become more significant as the responsibility now lies with the caterer for planning nutritionally adequate meals. Menu planning is the key to overcoming this problem.

Definition: Menu planning is defined as a simple process which involves application of the knowledge of food, nutrients, food habits, and likes and dislikes to plan wholesome and attractive meals.

The caterer who is responsible for providing meals has to decide on various aspects, such as:

1. Menu
2. Serving size
3. Food cost
4. Suppliers and quantities to be purchased
5. Standardized recipes to be followed
6. Type of service
7. Meal timings
8. Clientele.

The aim of menu planning is to:

1. Meet the nutritional needs of the individuals who will be consuming the food
2. Plan meals within the food cost
3. Simplify purchase, preparation, and storage of meals
4. Provide attractive, appetizing meals with no monotony
5. Save time and money
6. Minimize overhead expenditure, i.e., fuel, electricity, water, labour.

Menu planning is the most important aspect of planning and organization in the food industry. It is an advance plan of a dietary pattern over a given period of time.

Menus are of the following types.

Banquet menus: These are special menus for banquets or functions.

Institutional menus: Hospital menus, boarding school menus, and industrial can-teen menus.

Menus may be cyclic which means they are compiled to cover a specified period of time.

The length of the cycle may vary and is decidedupon by the management. A number of menus are set and repeated. They are often modified to take into account variations which may arise for a number of reasons.

Factors Influencing Meal Planning

Many factors influence the acceptability of a meal. Customers select what appeals most to them from a menu card based on individual likes and dislikes, budget, popularity of items, etc. However, while planning meals the following factors need to be considered.

Nutritional Adequacy

The most important consideration in menu planning is to ensure that the meal fulfils the nutrient needs of the individual consuming the meal. For example, if the meal is planned for an industrial worker, it must meet the RDAs for that age group. Foods from all basic food groups should be included in each meal so that the meal is balanced and nutritionally adequate. Nutrient needs may be modified for hospital diets (therapeutic diets).

Economic Considerations

The spending power of the clientele has to be kept in mind and meals have to be planned within the budget. Low cost nutritious substitutes should be included in the menu to keep the costs low. The food cost should be maintained, if the organization has to run profitably.

Type of Food Service

Menus should be planned in relation to the type of food service, whether it is cafeteria, seated service, buffet, etc.

Equipment and Work Space

The menu should be planned keeping the available equipment and work space in mind. Deep freezers, refrigerators, grinders, dough kneaders, deep fat fryers, boilers, etc. should be adequate.

Leftover Food

An effective manager should consider as to how leftovers could be rotated to obtain maximum profit. Adequate storage space and hygienic standards should be ensured to minimize the risk of contamination and spoilage of food.

Food Habits

Food habits of the customer is another important criteria which needs to be considered as food served has to be acceptable to the customer. Special attention should be paid when a particular type of community is catered to. Religious considerations should be known to the meal planner.

Availability

Some fruits and vegetables are seasonal. During the season the cost is reasonable and quality is better. Today, practically all fruits and vegetables are available throughout the year because of advanced preservation technology. However, seasonal fruits and vegetables should be given preference. Regional availability influences menu planning. For example, fish and sea food is fresh and cheaper in coastal areas.

Meat Frequency and Pattern

The meal timings and number of meals consumed in a day, whether meals are packed or served at the table, also influences the selection of food items on the menu. The age, activity level, physiological state, work schedule, and economic factors need to be known before planning meals for institutional catering.

Variety

This is one of the most important considerations while planning meals. A variety of foods from the different food groups should be included. The term variety means

1. Variety in food ingredients
2. Variety in recipe
3. Method of cooking
4. Colour, texture, and flavour
5. Variety in presentation and garnish.

A meal should look attractive and be appetizing. A judicious blend of flavours, attractive colour combinations, and different textures make food enjoyable and interesting. The method of cooking used for different items on the menu should vary.

All animal proteins Pulses, nuts, and oilseeds.

Protein, vitamin and mineral rich Protein, vitamin, mineral, fibre, oils.

For example, two deep fried items would make the meal heavy. Simple processes such as fermentation and sprouting not only contribute to improved flavour and digestibility, but also enhance the nutritive value of the meal.

A well planned meal which is nutritionally adequate would have a good satiety value and prevent the occurence of hunger-pangs before it is time for the next meal. The nutritional adequacy of a meal in an a la carte service depends on the food choices made by the customer. It is the duty of the caterer to offer adequate, nutrient dense foods to the clients, to choose from.

Planning Balanced Meals

Meal planning involves proper selection of food to ensure balanced meals. We have also read that food can be classified on the basis of its source, the nutrients present in it, or on the basis of its functions into 3-11 food groups. These food groups help us in planning balanced meals which supply all essential nutrients. In this chapter we will study the three basic food groups classified on the basis of functions performed by nutrients as this is the simplest way to ensure adequate nourishment to the body.

The three main functions performed by food are:

1. Providing energy
2. Body building and maintenance
3. Regulation of body processes and protection against infection.

On the basis of functions performed, food is classified into the following three groups.

1. Protective/regulatory foods
2. Body building foods
3. Energy giving foods.

Steps In Planning Balanced Meals

1. Collect information regarding the customer with respect to
   - Age
   - Gender
   - Activity level
   - Religion
   - Socioeconomic background
   - Food habits
2. Check the RDAs for energy and proteins
3. Prepare a food plan, i.e., list number of servings from each food group to meet the RDA
4. Decide on number of meals
5. Distribute servings for each meal
6. Select foods within each group and state their amount

7. Plan a menu
8. Cross check to ensure that all food groups are included in requisite amounts.

Using the above steps, plan a balanced diet for a day.

Example: Planning balanced meals for college students residing in a hostel.

Calculating the Nutritive Value of a Recipe

Using the food composition table, follow the steps given here to calculate nutritive value of a recipe.

1. List the ingredients used and their quantities in the recipe.
2. Prepare a table with the following blank columns, and fill up the ingredient and quantity column from the recipe.
3. Refer to the food composition tables for the nutrients present in 100 g of edible portion of each ingredient.
4. Calculate the nutrients present for the quantities used in the recipe.
5. Weigh the finished product to know the total yeild.
6. Divide these values by the number of portions to know the nutritive value per portion.
7. For general calculations, do not include salt, spices, baking powder, stock, or ingredients which are used in very small quantities (less than 10 g) except for sugar and fat. Baking powder and salt are calculated for their sodium content, and not for their proximate principles, for sodium-restricted diets only.

Special Nutritional Requirements

Pregnancy

Every infant should have the right to begin life with a full-term healthy body and receive the advantage of mother's milk. For this to happen, every would-be mother should take special care of her nutritional needs. Pregnancy is a period of remark-able anabolic activity. A 3.2 kg infant develops in 9 months of pregnancy from the nourishment received from the mother. If the diet is deficient during these months, the foetus draws upon the maternal reserves. A poor diet will ultimately affect both the infant and the mother, and may lead to complications of pregnancy such as pre-mature birth and low birth weight.

Weight gain: A healthy woman is expected to gain 16-24 lbs in the 3 trimesters of pregnancy of which the full-term infant weighs 7 lbs. The remaining weight is due to increase in tissues, an expanding blood volume, and energy stored in the form of fat as a calorie reserve.

Energy: The energy needs increase during the second and third trimesters of pregnancy. An additional 300 kcalories take care of the increased demands of pregnancy. Calorific value of food is adjusted according to the weight gained.

Protein: The protein requirement increases by 15 g for the synthesis of foetal and maternal tissues.

Minerals: The calcium requirement is 1,000 mg and iron is 38 mg for the formation of foetal bones and teeth, and formation of blood respectively. The absorption of minerals improves because of the increased requirement. Iodine in the form of iodized salt, helps in protecting the mother and child against goitre and cretinism.

Vitamins: With an increase in calories, the need for vitamins B_1, B_2, and niacin increases as they are needed for release of energy from carbohydrates, fats, and proteins. The need for folic acid and vitamin D also increases.

Nutrient dense foods should be selected to meet the extra demand for proteins, calcium, iron, and vitamin D and B-complex vitamins.

Lactation

Mother's milk is the most nutritious food designed for an infant. The nutritional needs during lactation are greater than the needs during pregnancy as the mother's body has to supply all nourishment to the rapidly growing infant. The conversion of dietary protein to milk protein is only 50% which means 2 g of good quality pro-tein is converted to 1 g of milk protein.

Energy: Calorie requirements increase by 550 kcal during the first six months of lactation, followed by a marginal decrease during the next six months with 400 kcal being adequate to meet the additional demands.

Proteins: An additional intake of 25 g in the first six months and 18 g in the next six months is adequate.

Minerals: Calcium requirement is 1,000 mg which is necessary for synthesis of milk. Additional iron is not prescribed as milk is a poor source of iron.

Fluid: The intake of fluid increases during lactation.

The diet should be nutritious, easy to digest with restrictions on strongly flavoured vegetables and spicy food.

Human milk is the natural food for the infant. It is safe and convenient as it does not involve sterilizing bottles and preparing formulas throughout the day. At the same time, it is easy to assimilate, has the correct temperature, and gives a safe and secure feeling to the infant, and sense of satisfaction to the mother.

During the first few days after delivery, colostrum is secreted which is not mature milk but a substance richer in protein and vitamin A. Colostrum is secreted in small quantities but is valuable as it increases the resistance to certain infections during the first few months of life.

The intervals of feeding should be fairly flexible instead of following a rigid schedule or a self-demand schedule. For premature and low birth weight (weight less than 2.5 kg) babies a fixed schedule is preferable. Milk is deficient in iron, vitamin C, and vitamin D. The baby is born with stores of these nutrients which suffice for 3 months. From the third month onwards, supplements should be gradually added to provide those nutrients which are not sup-plied by milk.

From the fifth to ninth month onwards, the infant should be weaned by substituting a cup feeding for a breastfeeding. This change should be gradual and at intervals till the infant is weaned from the breast to the cup before the age of one year.

Nutritional requirements: The next best substitute is cow's milk or modified buffalo milk. Buffalo milk is modified as it con-tains a larger percentage of protein, fat, calcium, and energy as compared to human milk and a lesser percentage of sugar. It is modified by partially skimming it, diluting it with water, and adding little sugar so that its composition resembles human milk.

Supplementary foods during the first year: The age at which supplementary foods should be included depends on a number of factors such as literacy of mother, economic factors, and time available. Supplements if not prepared, stored, and fed to the infant in hygienic conditions is one of the main causes for diarrhoea and gastrointestinal upsets. While introducing a new food the following points should be borne in mind:

1. Introduce one food at a time till the infant's system is used to the new food. Season with salt
2. Give a teaspoonful in the beginning

3. Never force-feed an infant
4. Start with a thin smooth consistency initially
5. Once the infant is used to supplements, introduce variety
6. Follow hygienic practices and do not give any leftover food or drink to an infant.

Choice of food supplements for infants:

1. Egg yolk
2. Strained cereals
3. Strained soup
4. Soft khichdi
5. Rice and dal
6. Pureed vegetables
7. Ready to eat infant cereals
8. Stewed fruit
9. Fruit juice.

Childhood

During these stages rapid growth takes place. There is an increase in height and weight because of increase in the bone and muscle mass.

Children use up a lot of energy in playing. They normally carry their lunch, which should be nutritious and interesting, or monotonous, in which case the packed lunch box comes back home unfinished or unopened. Special efforts should be made so that they do not meet their energy needs from junk food.

More energy per kg body weight is required for:

1. Rapid growth which takes place in this age group
2. Enhanced physical activity
3. High BMR as compared to an adult.

Proteins: Good quality proteins should be included to take care of body building and maintenance of tissues. Milk proteins are complete proteins and provide calcium as well.

Carbohydrates and fats: They provide calories, and spare proteins from being oxidized for energy. Refined carbohydrates and poor dental hygiene are the reasons for dental caries in children. The consumption of fruits and vegetables, both cooked and uncooked, should be encouraged. Junk food and aerated beverages should be discouraged.

Vitamins and minerals: If the diet is well planned keeping the principles of menu planning in mind, supplements may not be necessary.

Fluid:Adequate fluid is necessary, specially in active children who play outdoor games and sweat a lot. Fluids should not kill the appetite, which is seen when children are thirsty and drink a lot of liquid just before a meal.

Points to be considered:

1. Regular meal timings
2. No nibbling low nutrient dense snacks between meals
3. Do not force feed, let the normal appetite return
4. Meal timing should be pleasant
5. Food should be appetizing and attractively served. Finger foods should be preferred for children
6. Packed meals should not be messy to eat
7. Attractive colours and shapes appeal to children, and with a little imagination eating green vegetables and drinking milk could become more interesting.

Adolescence

Adolescence is a period of physiological stress for the body because of extremely rapid rate of growth. The appearance of sex characteristic is also accompanied by mental and emotional changes. The diet plays a crucial role in promoting growth, hence the RDAs for all nutrients are high during 13-18 years of age. Nutrients of particular importance are carbohydrates and fats for energy and proteins, iron and calcium for body building. During adolescence the basal metabolic rate (BMR) accelerates once again because of the growth spurt and other hormonal changes which take place. The BMR of boys is higher than that of girls because of more muscle tissue. Food habits change drastically because of peer pressure, maintaining one's figure and weight, skin problems, and the newly found independence.

Energy: Energy needs are high because of higher physical activity and BMR. Sports and aerobic exercises increase the energy requirements for sports people.

Proteins: A generous intake of high quality proteins is necessary for increase in muscles mass and skeletal development. A deficiency of calories and proteins can affect one's optimum height increase and resistance to infections.

Girls grow rapidly between 11-14 years and boys grow rapidly between 13-16 years. Initially, girls are taller than boys between 11-12 years. But later the growth rate is more in boys. Weight gain is a common problem in this age group which is tackled differently by both the sexes. Boys prefer exercise as a means of losing weight and body building while girls prefer going on a diet.

A deficiency of calcium/vitamin D during infancy or childhood results in rickets (deforming of bones). The weak bones cannot withstand the weight of the body and bend causing bow legs or knock knees. Once malformed, bones cannot be straightened. The effect of rickets is seen in adolescents and adults. Along with weight, emotional stress has an adverse effect on health and diet. Friction between parents and adolescents is another area of tension.

Snacks are the all-time favourite food, which often forms a large percentage of energy intake, favourite snacks being burgers, pizzas, pastries, ice creams, french fries, pav wada, pav bhaji, popcorn, and South Indian snacks. Milk is substituted by tea/coffee (which is more of a social need) or aerated beverages. This trend indicates a high consumption of refined flour, hydrogenated fats, and potato. South Indian snacks are a better option as they are made from a cereal-pulse combination. In general, the diet is deficient in the protein food group and protective food group. Food such as milk, green leafy vegetables, yellow fruits and vegetables, citrus fruits, and whole grain cereals are deficient in the diet of adolescents. Deficiencies during infancy and childhood may leave their scar in adolescence.

Old Age

Anorexia nervosa: This is a disorder seen in adolescent girls concerning figure and weight control. The onset of puberty results in widening of the pelvis and deposition of subcutaneous fat creating a psychological problem in figure conscious young girls. They resort to crash diets for weight loss. This self-induced starvation to attain a slim figure leads to weight loss, loss of muscle tissue, a low BMR, and serious health problems.

Bulimia: It is another disorder in which the appetite increases drastically followed by induced vomiting to throw out whatever has been consumed. Bulimia creates both emotional and physical problems and is also known as the gorging-purging syndrome. Both disorders are mainly seen in adolescent girls from affluent families.

Nutrition in old age or geriatric nutrition is gaining importance as longevity has increased and 65 plus is the fastest growing segment of our population. Aging is an individual phenomenon depending to a large extent on the health and nutrition status of an individual throughout his life span. People grow old at different rates and in different ways. Aging continues from birth throughout life.

Apart from the biological changes which occur with age, psychological and social changes are also seen. The body functions slow down, there is increasing social stress, and isolation. The elderly are often lonely, restless, and unhappy. They feel they are a burden and cannot contribute physically or monetarily to the family.

The following changes are seen in old age.

Body composition: As age advance, adults lose muscle mass and gain fat. Since active muscle is lost, BMR reduces with age. Bone loss occurs leading to a reduction in height. Collagen content increases giving rise to loss of elasticity in blood vessels, joints, and skin, seen as easy bruises, painful joints, and a wrinkled skin.

Sensory changes: The five senses of sight, hearing, taste, smell, and touch become dim with age. Old people cannot see without bright light, hearing is impaired affecting their social behaviour, and the other senses are less acute.

Digestive functions: Loss of teeth and ill-fitting dentures make it necessary to serve soft well-cooked food. Decreased secretion of saliva makes food difficult to swallow. Heartburn, diarrhoea, or constipation are other complaints.

Metabolic changes: Changes in metabolism, specially glucose tolerance and anaemias, are sometimes seen.

Osteopenia and osteoporosis are seen resulting in excessive bone loss, and fractures due to normal stress is another common finding. Because of age-related disabilities, the elderly should not live alone as they require assistance for all activities and company to overcome psychosocial changes in their life.

Nutritional requirements: The main features of the diet are as follows:

1. A low calorie diet of 1,000-1,500 kCal, as there is a reduction in the basal metabolic rate (BMR).
2. Good quality protein, calcium, and iron. Vitamin A, vitamin C, and folic acid to prevent anaemia.
3. Food should be easy to chew, liquid preparations should be included in each meal as they are light and easy to digest.
4. Hot meals should be served attractively as appetite is low.

5. Five small meals should be preferred to a three meal pattern.
6. Gas forming foods and strongly flavoured vegetables such as bengal gram, cauliflower, cabbage, and onion should be avoided.
7. Fibre rich foods such as whole grains, green leafy vegetables, pulses, and fruits are nutritious and provide fibre necessary for normal elimination. They should be incorporated into the diet gradually to prevent gastric irritation and gas.
8. Fluids: Ample fluids are necessary to flush out toxic wastes. The kidneys function at a slower pace with age, but fluids help in excretion of toxins.
9. Low calorie density and high satiety value foods should be included.

Effect of Quantity Cooking and Processing on Nutrients

Almost all foods consumed today need some form of cooking and processing before it is fit for service and consumption. Fruits and vegetables used in salads or for chutney are consumed uncooked. The nutrients we receive from the meals we consume depend to a large extent on cooking and processing practices which are being used. While some amount of nutrient loss is inevitable, cooking has many benefits which are listed below.

Benefits of Cooking Food

1. Cooking increases palatability
2. Cooking makes food easier to digest by destroying anti-digestive factors such as trypsin inhibitor in soya beans
3. Pathogenic microorganisms are destroyed
4. Shelf life is increased by destruction of spoilage organisms and denaturation of enzymes
5. The appearance of food improves, e.g., cooked meat versus raw meat.

Common Food Processing Techniques

Food prepared in large quantity in Institutional Kitchens or in food processing plants is more prone to loss of nutrients, if adequate care is not taken to retain or preserve the nutrient.

This is because if food is cooked in bulk, the pre-preparation begins hours in advance, for example, vegetables have to be cut in advance and if these are not blanched and refrigerated to inactivate enzymes, oxidative losses of labile vitamins will continue at room temperature. Apart from nutritive value, the crisp texture of salads is also lost and phenol containing vegetables will discolour and turn brown, making the dish unattractive and unappetizing.

Effect of heat on nutrients: Cooking has beneficial effects on carbohydrates because of gelatinization of starch, favourable browning reactions such as Maillard reaction and caramelization of sugar which gives colour and flavour to food. Proteins too take part in Maillard reaction along with sugar. Enzymes which catalyse undesirable enzymatic reactions in fruits such as apple and pears, and vegetables such as potato and brinjal are inactivated on blanching or cooking these foods. Enzymes which hasten oxidative destruction of vitamin C or ascorbic acid are denatured by blanching. Proteins get denatured by heat. The chemical reactions that take place when oil is heated continuously during deep fat frying bring about hydrolysis, oxidation, and polymerization of the oil.

Triglyceride

The release of moisture, high frying temperatures of 160°-190°C, presence of carbonized crumbs in the oil, and oxygen from the atmosphere during frying brings about oxidation of the oil. Repeated use of the frying medium forms thermal and oxidative products which can cause gastrointestinal irritation and destruction of vitamins. These products undergo polymerization and increase the viscosity of the oil. The oil darkens in colour, has a lower smoke point, and foams when used for frying. Such oil should be discarded. Fat soluble vitamins dissolve in fat used for deep frying specially if food to be fried is not well coated.

Effect of alkali: Alkali is used during cooking and processing to soften vegetables, make pectin soluble, and dissolve hemicellulose. It is also used as lye,sodium hydroxide) to peel vegetables during processing. A pinch of sodium bicarbonate added to green vegetables helps in brightening the green colour. However, B-complex vitamins and ascorbic acid are destroyed in an alkaline medium. The use of alkali to hasten the cooking process for vegetables and pulses should be discouraged. Excessive cooking in an alkaline medium not only destroys vitamins, but makes the texture mushy and gives a soapy taste to the product.

Effect of acid: An acidic medium while cooking helps preserve water soluble vitamins and retards enzymatic browning of certain fruits and vegetables. Vegetables and pulses take a longer time to cook in an acidic medium as acids precipitate pectin and hardens vegetables.

Effect of washing and soaking: While preparing food, water soluble vitamins and minerals leach out into the cooking or washing water. These losses can be minimized by washing the uncut fruit or vegetables and not soaking the cut vegetable in water. Soaking grains or pulses is beneficial as soaking increases digestibility and reduces cooking time.

Effect of sprouting and fermentation: Soaking whole grains overnight in water and tying them in a muslin cloth to allow them to germinate has many beneficial effects.

1. In sprouted grains, the dormant seed becomes active and synthesizes vitamin C.
2. Partial breakdown of carbohydrates, proteins, and fats begin, making it easier to digest.
3. The bioavailability of nutrients especially calcium and iron increases.
4. The active seeds synthesizes vitamin C or ascorbic acid and thiamine, riboflavin, and niacin content increases.

Exposure to air or oxidation: Exposure of finely divided foods to oxygen of the air reduces the vitamin C content by oxidation. The enzyme ascorbic acid oxidase is released when fruits and vegetables are cut. The enzyme activity is temperature dependant and can be inactivated by blanching, or by storing cut fruits and vegetables at refrigeration temperatures or by adding acid.

Vitamin A is destroyed on exposure to air. The colour of cut carrots (carotene) fades due to oxidation and B-complex vitamins are also affected.

Milling: Whole meal flour contains all nutrients present in the grain. In flour with 100% extraction no nutrients are lost. Low extraction flours (45% extraction) are light in colour and are mainly starch with some protein and fat. Approximately 70°/0 of all B-complex vitamins, minerals, and dietary fibre present in the whole grain are lost during milling. Polished rice (the form in which rice is consumed) loses 75% vitamin B_1 or thiamine, while parboiling helps in retaining some of the vitamin.

Cooking and processing practices vary widely from one region to another, hence no authentic information on exact losses can be known. While cooking has both adverse and beneficial effects, proper practices can minimize the adverse effects and maximize the benefits so that food can become more whole-some and safe.

Chapter 2

Operations of Food Services

Kitchen work and chefs The heart of the hotel is the kitchen. Guests who patronize these hotels would visit them for various reasons, for locality, comfort, pricing and even for the food that is served and presented to them. Also there are some who visit these hotels for the purpose of having a meal or snack. Whatever the case maybe, the kitchen plays an important role for the existence and sustenance of the hotel.

Working in a hotel kitchen could be interesting, tiring, challenging and even exciting. With all the modern technology and gadgetry, these kitchens have turned out be somewhat high tech and more convenient. Some kitchens in luxurious hotels are so well equipped and maintained, with freezer rooms, to separate kitchens for the preparation of cakes, bread and pastry. With all the novel stoves, ovens and microwaves it is indeed a fantastic place to work in and a wonderful experience. Depending on size and class of the hotel, the kitchen staff may vary. Most of the top hotels have an Executive Chef and working under him there are different categories of chefs.

They are Chef de Cuisine, Chef de Partie, Demi Chef, Commis Chef, Sous Chef etc. The duties of these chefs will vary from one to another. The Executive Chef or the Head Chef generally will oversee and supervise the department, and will instruct them accordingly as to what the duties for the day are. They are also responsible in all matters pertaining to the department, such as leave, attending to personal issues, attending board meetings and scheduling the work to each staff member. They are expected to work on shift basis as a hotel function 24-7. Not only preparation, the presentation of the food is also equally important. An expected attribute of a chef is creativity and novelty. They are responsible from the start of the preparation of the food to the final presentation. They are expected to present the food to the guest in a manner that is appealing. Most of the elite hotels are continuously trying to improvise ways and means to allure guest to their hotels by having food fares and promotions. This has been a successful way earning revenue and is patronized by many guests for food experiences. Being creative could help a chef to present a meal in the most appetizing way, thus winning the heart of the customer.

Good communication skills are important to be successful in your mission. While constant training of the staff is done by the Executive Chefs, they are also expected at times to speak to guests and attend to some special need. Cleanliness and good hygiene is maintained at all times, it is said that the standard of the hotel could be judged by its kitchen.

Food and Beverage Services

Food and beverage services are among the most visible locations at ski areas, and a ski area's environmental commitment is often judged by the environmental practices within these establishments. Pollution prevention (P2) opportunities for restaurants address a broad range of operations, including solid waste management, grounds and facility maintenance, and restaurant supply purchasing.

A restaurant's pre-tax profit is typically only 3 to 9 percent of its total revenue; therefore, money saved through reductions in operating costs (that is, through reduced energy consumption and water use) can significantly increase the profit margin. While some restaurants have already taken advantage of the many P2 opportunities available to them, a survey by the National Restaurant Association shows that for some of the most common P2 techniques and best practices, there is still room for improvement across the industry.

Numerous programs exist for the food and beverage service industry that help restaurants and their customers minimize environmental impacts. This chapter uses the Green Restaurant Association's approach as an outline to present the environmental practices a restaurant should consider.

Green Restaurant Association

Involvement in environmental programs is a good way for a restaurant to learn more about environmental best practice opportunities. An environmental resource for the food service industry is the Green Restaurant Association (GRA), which helps member restaurants reduce environmental impacts with twelve "eco-steps".

The GRA's primary operational components are research, environmental consulting, education, public relations and marketing, and community organizing and consumer activism.

To become a member of the GRA, a restaurant must sign a statement of its commitment to making environmental improvements based on the 12 "eco-steps" listed below.

After a member restaurant has made positive environmental changes, the GRA will include the restaurant in the "Green Restaurant Guide," which serves as a directory for environmentally conscious restaurants. The GRA also offers fee-based consulting services to member restaurants and initiates public relations and marketing initiatives, such as having restaurants featured on CNN, to increase restaurant exposure and consumer awareness.

GRA's 12 eco-steps are outlined below and discussed further in Sections 9.2 through 9.13.

1. Elimination of Polystyrene Foam (commonly known as styrofoam). Replace all polystyrene foam products with environmentally friendly alternatives: paper, bamboo or sugarcane paper, recyclable plastic, biodegradable plastic, and so on. *This is the minimum environmental standard for becoming a member of the GRA.*
2. Comprehensive Recycling. Initiate or improve recycling programs for glass, plastic, bimetal, cardboard, and mixed paper.
3. Waste Reduction and Reuse. Increase bulk purchasing and reduce excessive packaging for food, condiments, and so on. Replace disposable products with reusable alternatives: eating ware, aprons, tea strainers, cups, and so on.
4. Biodegradable Plastic. Transition to corn-based "plastic" products that are biodegradable and petroleum-free: cups, utensils, garbage bags, and straws.
5. Recycled Products. Transition to recycled products with the highest postconsumer content available and non-tree-fiber paper products: napkins, paper towels, toilet paper, office paper, take-out containers, coffee jackets, plates, and bowls.
6. Non-Chlorine-Bleached Paper Products. Transition to non-bleached or non-chlorinebleached paper products: cups, wax

paper, plates, take-out containers, bags, pastry bags and grabbers, napkins, paper towels, coffee filters, and office paper.

7. Nontoxic Cleaners, Landscaping and Pest Management. Replace hazardous chemical products with biodegradable and nontoxic alternatives; dish detergent, germicides, disinfectants, toilet bowl cleaners, drain cleaner, floor wash, floor polish, glass cleaners, degreasers, and laundry detergent. For landscaping, switch to nontoxic, nonsynthetic, and organic fertilizers, pesticides, or herbicides. For pest control, use nontoxic products or services.
8. Energy Efficiency. Improve the energy efficiency of lighting, refrigeration, air conditioning, gas appliances, and so on. Obtain assistance with using government and private rebate programs. Make connections with energy consultants to learn about more extensive programs.
9. Water Efficiency. Improve the water efficiency of toilets, faucets, laundry, sprinkler systems, and so on.
10. Composting. Divert food waste from landfills and create nutrient-rich soil for gardening and landscaping.
11. "Green" Electricity. Change to an energy provider that uses solar, wind, small-scale hydroelectric, geothermal, or methane-based power that is renewable and less polluting to ecosystems.
12. Employee Education Program. Train all employees, managers, and owners. Topics covered should include
    - An environmental profile of the restaurant industry
    - A history of environmental issues relevant to food service: landfills, water pollution, air pollution, clear-cutting, and global warming
    - Data describing the restaurant impacts (positive and negative) on the environment.

Elimination of Polystyrene Foam

Polystyrene foam (Styrofoam) is widely used in restaurants in insulated cups for hot beverages and take-out containers. Production of polystyrene involves use of known (benzene) and suspected (styrene and 1,3-butadiene) human carcinogenic substances (styrene and 1,3-butadiene).

Styrene is also known to be toxic to the reproductive system. Polystyrene can be recycled; however, its recycling rates are low.

Styrofoam is light in weight, but bulky in size, so hauling Styrofoam to the nearest available recycling facility is often not economically feasible. The value of a load of Styrofoam may not even cover the cost of shipping. To reduce environmental impacts, restaurants should consider switching from Styrofoam to non-bleached paper wraps, cardboard containers, or other sustainable food packaging.

According to a report published by the Environmental Defence Waste Reduction Task Force, McDonalds has completed the switch from polystyrene foam "clamshells" to paper-based wraps for packaging its sandwich items. The wraps provide a 70 to 90 percent reduction in packaging volume, resulting in significantly less space being consumed in landfills. Compared to the polystyrene foam boxes they replaced, the new sandwich wraps also offer a substantial savings in energy used and substantial reductions in pollutant releases measured over the full life-cycle of the packaging. McDonalds is also testing Earth Shell packaging products in 300 of its restaurants throughout the U.S.. These sustainable containers are made from potato starch, natural limestone, 100 percent postconsumer recycled fiber, biodegradable polymer and wax coatings, and water.

Comprehensive Recycling

The "Recycling Guidebook for the Hospitality and Restaurant Industry" provides a general overview of developing and implementing a comprehensive recycling program. The first step mentioned in this resource is to conduct a waste audit in order to evaluate the waste stream, enabling a restaurant to better target the commodities that should be included in a recycling program. Materials that are commonly recycled in restaurants include;

- *Paper*, including cardboard, computer paper, register tape, and telephone books
- *Metals*, including aluminum, tin, and steel cans
- *Green, brown, and clear glass*
- *#1 Polyethylene* (PET) and *#2 high-density polyethylene (HDPE)* plastics
- *Printer cartridges.*

Convenience is the key to a successful recycling program. An EPA Waste Wise tip sheet notes that a convenient collection system will encourage both customers and employees to carefully sort recyclables by material type and to eliminate contaminants. Collecting uncontaminated recyclables (commodities that are properly sorted and free of excess food and beverage waste) will save time otherwise spent in sorting out contaminants. Further, uncontaminated recyclables have higher value if they are sold. Presented below are other restaurant recycling tips listed in the "Recycling Guidebook for the Hospitality and Restaurant Industry."

- Recycling and trash bins should look different from each other and be clearly marked. Both types of bins should be conveniently located in the kitchen and bar areas so that employees will use them.
- Self-serve establishments should post signs to inform customers about the recycling program and provide specific instructions. Either strategically places bins for collection of recyclable commodities, or have customers leave such commodities on a designated counter for collection by staff.
- If lack of space is a problem, specially designed equipment such as can, glass, and plastic crushers are available to reduce the volume of recyclable materials.
- Recyclable collection bins in public areas should be well-marked. Choose bins with specialized openings, such as a hole for cans or a slot for newspapers, for these areas.
- Set up a logbook or a receipt system to record the volume of recyclables leaving the premises in order to facilitate tracking and compensation.

Waste Reduction and Reuse

Waste reduction and material reuse should be the first step in minimizing the waste that a restaurant produces because it is more efficient to reduce waste at the source or reuse material than it is to recycle. As mentioned in audit is an effective way to evaluate a restaurant's waste streams. Doing so enables staff to identify wastes that are nonessential to operations, such as excess packaging material. This process can also identify disposable products that can be replaced with durable, reusable ones. For example, disposable plastic tableware and silverware should be replaced with washable or compostable utensils.

EPA's Waste Wise program is a free, 3-year, goal-oriented program that assists businesses in assessing and reducing their waste streams.11 EPA designed Waste Wise to be a flexible program, in which the participant determines how much time and money to invest. Key aspects of successful Waste Wise programs include;

- Waste assessments
- Employee education
- Measurement and reporting
- Program maintenance.

The above aspects are fundamental in assessing, developing, implementing, and maintaining a viable waste reduction and reuse program. Waste assessments provide a full understanding of waste streams and provide a basis for targeting specific waste reduction goals. However, these goals will only be reached with the full support of the restaurant – from managers and servers to cooks and food preparation staff. Therefore, employee education is critical to the success of the program. It is also important that program efforts be measured regularly.

Tracking and documenting the costs, savings, and effects of a waste reduction and reuse program are the only ways to determine the environmental and economic impacts of implementing the program. Program data can also provide insight into the effectiveness of employee education programs. Lastly, program maintenance is imperative to realize long-term benefits from a waste reduction and reuse program. A successful program is not a result of one-time changes but a cumulative result of permanent procedural and behavioural changes. Several other guidance documents provide specific waste reduction and reuse ideas for the restaurant industry. Presented below are some examples of common waste reduction techniques that are discussed in "Food for Thought: Restaurant Guide to Waste Reduction and Recycling."

- Purchase and serve beverages in bulk rather than in bottles, cans, or individual packets.
- Buy bar mixes, juices, and coffee in bulk.
- Use health department-approved, refillable condiment containers.
- Purchase cleaning supplies in concentrated form.
- Use cloth towels and reusable table linens and tableware.
- Use vendors that take back packaging material and pallets.

Another way that restaurants can minimize their waste and overall environmental impact is purchasing products through local food supply vendors and grocers. Local sourcing is of particular benefit to the tourism sector, where visitors appreciate locally distinctive food, drink, and other products. For example, several Aspen Skiing

Company-owned restaurants purchase natural, chemical-free beef and hamburger from local ranchers to support the ranching community and thus preserve open space, as well as to provide guests with healthy food. The pilot program began in winter 1998 and expanded significantly in 2000. The beef costs twice as much per pound, however

Aspen Skiing Company absorbs the additional cost. By purchasing local produce and supplies, restaurants minimize the affects of excessive transportation and shipping. Such efforts should be accounted for when evaluating the environmental impact and overall performance of a restaurant.

Biodegradable Plastic

As long as customers wish to take their food "to go," restaurants will need to stock disposable goods such as food containers and silverware. According to Biocorp, a manufacturer of biodegradable plastic tableware and silverware, nearly 113 billion disposable cups, 39 billion disposable eating utensils, and 29 billion disposable plates are used in the U.S. every year, and half these items are made of plastic. Because restaurants cannot control what happens to these items once they are in customers' hands, the most effective way to minimize their environmental impacts is to purchase biodegradable products.

Biodegradable products are typically made of corn, starch, or paper with an easily biodegradable coating. The Earth Shell is one example of a biodegradable food service product. Biocorp also manufactures biodegradable composting bags, silverware, plates, and beverage containers.

"Scientists Perfecting Planet-Friendly Plastics," an article published by the Environmental News Network, explains that the starch used to create these biodegradable plastics– typically wheat gluten – costs about 15 cents per pound, whereas the least costly commercial plastics cost about $1 per pound. Thus, when this starch-based plastic becomes widely available to manufacturers, it could be the cheapest plastic available.

Currently, however, most manufactured environment-friendly plastics cost about $2.50 per pound, although recent projects have brought their costs down to approximately $1.50 per pound. Because biodegradable plastics are more expensive than regular plastic, the biodegradable plastic industry has a challenge in breaking into a very competitive market.

Recycled Products

Restaurants use several types of products that can be replaced with items made from recycled materials. Most of the products with recycled content that can easily be purchased are paper products such as napkins, toilet paper, tissues, paper towels, paper tablecloths, take-out containers, register tape, and office paper. Since 1998, EPA has required federal facilities to purchase products with recycledcontent material. The resulting increased demand for recycled content paper has driven down its cost to the point that recycled-content products are the same price or less expensive than virgin material products.

Non-chlorine-bleached Paper Products

Restaurants should avoid purchasing bleached paper products. Seventh Generation, a vendor of environmentally friendly, nontoxic consumer products, explains that chlorine is used by the paper industry for two purposes: to dissolve lignin, a natural material that holds a tree's cellulous fibers together, and to whiten the final paper product. However, chlorine reacts with the virgin natural substances of trees and recycled paper material to form both dioxins and organochlorines.

As these substances are discharged and accumulate in the environment, they can have profound long-and short-term health effects on exposed humans and wildlife. One concern regarding dioxins is that they are known carcinogens, or cancer-causing substances. Short-term reactions to overexposure to chlorine may also occur, including airway inflammation and bronchial hyperresponsiveness. Organochlorines are of concern because they are suspected to be endocrine modifiers, which act as hormones in the body and can disrupt the human immune system.

Nontoxic Cleaners, Landscaping, and Pest Management

Restaurant personnel use a wide variety of chemicals to clean kitchen, dishwashing, and restroom facilities. Some cleaning products used include toilet and tile cleaner, glass cleaner, carpet cleaner, spot remover, disinfectant, and oven cleaner. Many of these products contain chemicals that are harmful to human health and the environment, which is a concern for both restaurant staff and customers. Vendors offer environmentally preferable cleaning supplies with equal or better cleaning performance at equal or less cost. Many restaurants have outdoor areas with landscaping. There are several relatively simple steps that a restaurant manager can take to minimize the environmental impact of maintaining these areas, including water conservation and selection of climate-appropriate indigenous plants. Some "green" practices for landscape management include

- Watering vegetation using "deficit" irrigation, or frequent light watering
- Selecting plants based on watering needs (Typically, indigenous plants are most efficient in using water.)
- Watering grounds during the coolest hours of the day (typically at night) Finally, pest management can also be an issue for restaurants.

Energy Efficiency

Most restaurant operations, such as cooking and cleaning, are energy-intensive. EPA's EnergyStar® Program. estimates that reducing energy consumption by 20 percent can increase a restaurant's profit by one-third, a good economic incentive to make energy efficiency investments. Making this sort of reduction is feasible when restaurants implement strategic energy conservation measures. The EnergyStar® Program analysis of restaurant energy use shows that over 60 percent of a restaurant's energy is used for cooking, heating water, and heating the establishment.

The National Restaurant Association also published tips for reducing energy costs for restaurants.

Turn on and Off

- Preheat equipment in accordance with manufacturer specifications; post preheating times near equipment.
- Turn off appliances when not in use.
- Install occupancy sensors in walk-in refrigerators and freezers.

Cooking Efficiency – Lower it/Fill it up

- Cook using equipment at full capacity when possible.
- Cook at lowest temperatures first.
- Turn off equipment during downtime.
- Use lids to minimize heat loss.

Watch Thermostats – Stay Cool and Save Money

- Set thermostats to manufacturer-recommended temperatures.
- Apply new-generation "clear" coatings to reduce solar gain from large, south-and west-facing windows.
- Use the "unoccupied" and "night setback" thermostat options.

- Ensure tamper-proof temperature settings by using locking covers on thermostats.

Keep It Clean for Energy Efficiency

- Clean condenser and evaporator coils on air conditioning and refrigeration equipment.
- Change all filters regularly.
- Preventively maintain equipment.

Dishwashing Equipment

- Heat water only to the temperature required for specific tasks.
- Install equipment of proper size.
- Fully load the machine for each cycle.

As restaurants undergo renovation or technology upgrades, managers should evaluate the energy efficiency of new equipment. Green Seal is an independent, nonprofit organization dedicated to protecting the environment by promoting the manufacture and sale of environmentally responsible consumer products.

Green Seal sets environmental standards and awards a "Green Seal of Approva" to products that cause less harm to the environment than other, similar products. Commercial consumers can use Green Seal's web site to search for resourceefficient Products

Water Efficiency

Restaurants use water in almost every aspect of their operations, including food preparation and cooking, cleaning, in restroom facilities, and as a beverage for customers. For each of these aspects, there are technologies or best management practices (BMP) that conserve water. In a collaborative effort, Agricultural and Biological Engineering, EPA, and Purdue University created an environmental enrichment toolkit for the lodging industry. Included in this document is a section that outlines the following water conservation tactics for restaurants:

- Wash food products in buckets, bowls, or other containers.
- Only run dishwashers with full loads.
- Regularly inspect dishwasher pumps for water leaks.
- Defrost or thaw frozen foods in the refrigerator instead of water.
- Install low-flow taps in kitchens and restrooms.
- Use low-flow toilets in restrooms.

- Immediately fix any leaking or dripping faucet.
- Install infrared-activated faucets and toilets in restrooms.
- Purchase and use water-saving kitchen equipment.
- Track water consumption by regularly monitoring utility bills.
- Establish an effective employee training program on water conservation.

Composting

EPA estimates that food wastes comprise 6.7 percent by weight of the total U.S. municipal solid waste stream. One option for diverting food wastes from landfills is composting. Businesses with well-established composting programs divert 50 to 100 percent of their food scraps and reduce their overall solid waste by 33 to 85 percent.

Composting can be done both on and off site; however, for restaurants where space is limited, the most feasible option is to collect food scraps for an off-site composting program. This option also transfers the responsibility of monitoring the chemical balance of the compost from the restaurant owner to the local composting operation.

However, technologies are available that enable restaurants to manage their compost on site.

Case Study: Composting At Keystone Ranch

Keystone Ranch, a horse ranch resort in Keystone, Colorado, implemented a composting program that combines horse manure with vegetable and meat scraps from the Ranch Restaurant.

The ranch rents a 30-cubic yard container from a local hauler that holds almost 20 tons of material. Biodegradable cornstarch-based bags are placed at four stations to collect food waste during food preparation and uneaten food from customers. Keystone Ranch estimates that between 10 and 15 percent of the contents of the full 30-cubic yard container is food waste. The rest of the compost mixture is manure collected from the stalls of 15 to 80 horses.

The hauler transports the container to the Twin Landfill Corporation in a nearby town, where the compost is further processed, screened, and then sold to landscapers and farmers. As a result of the composting program, which complements an established glass, aluminum, tin, cardboard, and paper recycling program, the Ranch Restaurant is "rapidly approaching a zero-waste status." Because the tipping fee for compostable material is 30 percent less than that for solid waste, the program sustains itself financially and has diverted 450 tons of organic waste from the local landfill.

Green Electricity

In addition to efforts to minimize their energy use, restaurants can minimize the environmental impacts of the energy they do use by purchasing renewable electricity. By using a renewable source of energy, a restaurant decreases its dependency on nonrenewable fossil fuels and eliminates greenhouse gas emissions associated with its electricity use. Renewable energy sources include wind, solar, geothermal, and biomass.

The availability of green electricity varies from region to region, so restaurants interested in purchasing energy from a renewable source should contact their city or state government agencies or local electric utility. Non-profit green energy certification programs, such as Green-e and Cleaner and Greener, may also be able to help direct restaurant owners to local green energy providers.

Employee Education Program

The development of an employee education program is critical to maintaining an environmentally responsible restaurant. An environmental education program should teach both the "how" and "why" of each aspect of the program. An understanding of why an employee should ensure that certain environmental procedures are followed adds meaning to the task. Because of high turnover in many restaurants, environmental education should be a regular, multimedia program, and should include both written and verbal instruction in all languages of the employees. Environmental awareness expectations should be outlined during new-hire training and reinforced throughout the period of employment.

Restaurant managers should consider acknowledging employees who demonstrate superior dedication to minimizing environmental impacts. For example, employees may be offered a percentage of the cost savings resulting from environmental projects implemented at their suggestion. Such recognition would not only reward the employees, but would send a message to customers that the restaurant is actively pursuing "greener" practices.

Once the expectation of excellent environmental performance is established, maintaining the program will require less effort. Finally, restaurant managers should encourage employees to provide feedback regarding areas for improvement. Many employees will have ideas of how to improve the environment they work in, and with a forum to voice ideas, innovations can be made.

Food and Beverage Serving and Related Workers

Nature of the Work

Food and beverage serving and related workers are the front line of customer service in full-service restaurants, casual dining eateries, and other food service establishments. These workers greet customers, escort them to seats and hand them menus, take food and drink orders, and serve food and beverages. They also answer questions, explain menu items and specials, and keep tables and dining areas clean and set for new diners. Most work as part of a team, helping coworkers to improve workflow and customer service.

Waiters and waitresses, also called *servers*, are the largest group of these workers. They take customers' orders, serve food and beverages, prepare itemized checks, and sometimes accept payment. Their specific duties vary considerably, depending on the establishment. In casual-dining restaurants serving routine, straightforward fare, such as salads, soups, and sandwiches, servers are expected to provide fast, efficient, and courteous service. In fine dining restaurants, where more complicated meals are prepared and often served over several courses, waiters and waitresses provide more formal service emphasizing personal, attentive treatment at a more leisurely pace.

Waiters and waitresses may meet with managers and chefs before each shift to discuss the menu and any new items or specials, review ingredients for potential food allergies, or talk about any food safety concerns. They also discuss coordination between the kitchen and the dining room and any customer service issues from the previous day or shift. In addition, waiters and waitresses usually check the identification of patrons to ensure they meet the minimum age requirement for the purchase of alcohol and tobacco products wherever those items are sold.

Waiters and waitresses sometimes perform the duties of other food and beverage service workers, including escorting guests to tables, serving customers seated at counters, clearing and setting up tables, or operating a cash register. However, full-service restaurants frequently hire other staff, such as hosts and hostesses, cashiers, or dining room attendants, to perform these duties.

Bartenders fill drink orders either taken directly from patrons at the bar or through waiters and waitresses who place drink orders for dining room customers. Bartenders check the identification of customers seated at the bar to ensure they meet the minimum age requirement for the purchase of alcohol and tobacco products. They prepare mixed drinks, serve bottled or draught beer, and pour wine or other beverages. Bartenders must know a wide range of drink recipes and be able to mix drinks accurately, quickly, and without waste. Some establishments, especially those with higher volume, use equipment that automatically measures, pours, and mixes drinks at the push of a button. Bartenders who use this equipment, however, still must work quickly to handle a large volume of drink orders and be familiar with the ingredients for special drink requests. Much of a bartender's work still must be done by hand.

Besides mixing and serving drinks, bartenders stock and prepare garnishes for drinks; maintain an adequate supply of ice, glasses, and other bar supplies; and keep the bar area clean for customers. They also may collect payment, operate the cash register, wash glassware and utensils, and serve food to customers who dine at the bar. Bartenders usually are responsible for ordering and maintaining an inventory of liquor, mixers, and other bar supplies.

Hosts and hostesses welcome guests and maintain reservation and waiting lists. They may direct patrons to coatrooms, restrooms, or to a place to wait until their table is ready. Hosts and hostesses assign guests to tables suitable for the size of their group, escort patrons to their seats, and provide menus. They also enter reservations, arrange parties, and assist with other special requests. In some restaurants, they act as cashiers.

Dining room and cafeteria attendants and bartender helpers—sometimes referred to collectively as the bus staff—assist waiters, waitresses, and bartenders by cleaning and setting tables, removing dirty dishes, and keeping serving areas stocked with supplies.

They may also assist waiters and waitresses by bringing meals out of the kitchen, distributing dishes to individual diners, filling water glasses, and delivering condiments. They may carry trays to dining tables for patrons. *Bartender helpers* keep bar equipment clean and glasses washed. *Dishwashers* clean dishes, cutlery, and kitchen utensils and equipment.

Food also is prepared and served in limited-service eateries, which don't employ servers and specialize in simpler preparations that often are made in advance. Two occupations with large numbers of workers are common in these types of establishments: *combined food preparation and serving workers, including fast food*; and *counter attendants, cafeteria, food concession, and coffee shop*. Combined food preparation and serving workers are employed primarily by fast food restaurants. They take food and beverage orders, retrieve items when ready, fill drink cups, and accept payment.

They also may heat food items and assemble salads and sandwiches, which constitutes food preparation. Counter attendants take orders and serve food in snack bars, cafeterias, movie theatres, and coffee shops over a counter or steam table. They may fill cups with coffee, soda, and other beverages and may prepare fountain specialties, such as milkshakes and ice cream sundaes. Counter attendants take carryout orders from diners and wrap or place items in containers.

They clean counters, write itemized bills, and sometimes accept payment. Other workers, referred to as *foodservers, nonrestaurant*, serve food to patrons outside of a restaurant environment. They might deliver room service meals in hotels or meals to hospital rooms or act as carhops, bringing orders to parked cars.

Work environment. Food and beverage service workers are on their feet most of the time and often carry heavy trays of food, dishes, and glassware. During busy dining periods, they are under pressure to serve customers quickly and efficiently. The work is relatively safe, but injuries from slips, cuts, and burns often result from hurrying or mishandling sharp tools. Three occupations—food servers, nonrestaurant; dining room and cafeteria attendants and bartender helpers; and dishwashers—reported higher incident rates than many occupations throughout the economy.

Part-time work is more common among food and beverage serving and related workers than among workers in almost any other occupation. In 2008, those on part-time schedules included half of all waiters and waitresses and almost three-fourths of all hosts and hostesses.

Food service and drinking establishments typically maintain long dining hours and offer flexible and varied work opportunities. Many food and beverage serving and related workers work evenings, weekends, and holidays. The long business hours allow for more flexible schedules that appeal to many teenagers who can gain valuable work experience. More than one-fifth of all food and beverage serving and related workers were 16 to 19 years old in 2008—about six times the proportion for all workers.

Training, Other Qualifications, and Advancement

Most food and beverage service jobs are entry level and require a high school diploma or less. Generally, training is received on the job; however, those who wish to work at more upscale restaurants, where income from tips is greater and service standards are higher, may need previous experience or vocational training.

Education and training. There are no specific educational requirements for most food and beverage service jobs. Many employers prefer to hire high school graduates for waiter and waitress, bartender, and host and hostess positions, but completion of high school usually is not required for fast-food workers, counter attendants, dishwashers, and dining room attendants and bartender helpers.

Many entrants to these jobs are in their late teens or early twenties and have a high school education or less. Usually, they have little or no work experience. Food and beverage service jobs are a major source of part-time employment for high school and college students, multiple job holders, and those seeking supplemental incomes.

All new employees receive some training from their employer. They learn safe food handling procedures and sanitation practices, for example. Some employers, particularly those in fast-food restaurants, teach new workers using self-study programs, on-line programs, audiovisual presentations, and instructional booklets that explain food preparation and service skills. But most food and beverage serving and related workers pick up their skills by observing and working with more experienced workers.

Some full-service restaurants also provide new dining room employees with some form of classroom training that alternates with periods of on-the-job work experience.These training programs communicate the operating philosophy of the restaurant, help establish a personal rapport with other staff, teach formal serving techniques, and instill a desire to work as a team. They also provide an opportunity to discuss customer service situations and the proper ways to handle unpleasant circumstances or unruly patrons.

Some food serving workers can acquire more skills by attending relevant classes offered by public or private vocational schools, restaurant associations, or large restaurant chains. Some bartenders acquire their skills through formal vocational training either by attending a school for bartending or a vocational and technical school where bartending classes are taught. These programs often include instruction on State and local laws and regulations, cocktail recipes, proper attire and conduct, and stocking a bar. Some of these schools help their graduates find jobs.

Although few employers require any minimum level of educational attainment, some specialized training is usually needed in food handling and legal issues surrounding serving alcoholic beverages. Employers are more likely to hire and promote employees based on people skills and personal qualities than education.

Other qualifications: Restaurants rely on good food and customer service to retain loyal customers and succeed in a competitive industry. Food and beverage serving and related workers who exhibit excellent personal qualities—such as a neat appearance, an ability to work as part of a team, and a natural rapport with customers—will be highly sought after. Most States require workers who serve alcoholic beverages to be at least 18 years of age, but some States require servers to be older. For bartender jobs, many employers prefer to hire people who are 25 or older. All servers that serve alcohol need to be familiar with State and local laws concerning the sale of alcoholic beverages.

Waiters and waitresses need a good memory to avoid confusing customers' orders and to recall faces, names, and preferences of frequent patrons. Knowledge of a foreign language can be helpful to communicate with a diverse clientele and staff.

Restaurants and hotels that have rigid table service standards often offer higher wages and have greater income potential from tips, but they may also have stiffer employment requirements, such as prior table service experience or higher education attainment than other establishments. Advancement. Due to the relatively small size of most food-serving establishments, opportunities for promotion are limited. After gaining experience, some dining room and cafeteria attendants and bartender helpers advance to waiter, waitress, or bartender jobs.

For waiters, waitresses, and bartenders, advancement usually is limited to finding a job in a busier or more expensive restaurant or bar where prospects for tip earnings are better. Some bartenders, hosts and hostesses, and waiters and waitresses advance to supervisory jobs, such as dining room supervisor, maitre d', assistant manager, or restaurant general manager.

A few bartenders open their own businesses. In larger restaurant chains, food and beverage service workers who excel often are invited to enter the company's formal management training program.

Employment

Food and beverage serving and related workers held 7.7 million jobs in 2008. The distribution of jobs among the various food and beverage serving occupations was as follows:

Combined food preparation and serving	
workers, including fast food	2,701,700
Waiters and waitresses	2,381,600
Counter attendants, cafeteria,	
food concession, and coffee shop	525,400
Dishwashers	522,900
Bartenders	508,700
Dining room and cafeteria attendants	
and bartender helpers	420,700
Hosts and hostesses, restaurant,	
lounge, and coffee shop	350,700
Food servers, nonrestaurant	189,800
All other food preparation and	
serving related workers	50,900

The overwhelming majority of jobs for food and beverage serving and related workers were found in food services and drinking places, such as restaurants, fast food outlets, bars, and catering or contract food service operations. Other jobs were in hotels, motels, and other traveller accommodation establishments; amusement, gambling, and recreation establishments; educational services; nursing care facilities; and civic and social organizations.

Jobs are located throughout the country but are more plentiful in larger cities and tourist areas. Vacation resorts offer seasonal employment.

Job Outlook

Average employment growth is expected, and job opportunities should be excellent for food and beverage serving and related workers as turnover is generally very high among these workers, but job competition is often keen for jobs at upscale restaurants.

Employment change. Overall employment of these workers is expected to increase by 10 percent over the 2008-18 decade, which is about as fast as the average for all occupations. Food and beverage serving and related workers are projected to have one of the largest numbers of new jobs arise, about 761,000, over this period. The growth in jobs is expected to increase as the population continues to expand. However, employment will grow more slowly than in the past as people change their dining habits. The growing popularity of take-out food and the growing number and variety of places that offer carryout options, including at many full-service restaurants, will slow the growth of waiters and waitresses and other serving workers.

Projected employment growth will vary by job type. Employment of combinedfood preparation and serving workers, which includes fast-food workers, is expected to increase faster than the average for all occupations. The limited service segment of the food services and drinking places industry has a low price advantage, fast service, and has been adding healthier foods. Slower than average employment growth is expected for waiters and waitresses, hosts and hostesses, and dining room and cafeteria attendants and bartender helpers, as more people use take-out service.

Employment of bartenders, dishwashers, and counter attendants, cafeteria, food concession, and coffee shop will grow about as fast as average. Nonrestaurant servers, such as those who deliver food trays in hotels, hospitals, residential care facilities, or catered events, are expected to have average employment growth.

Job prospects. Job opportunities at most eating and drinking places will be excellent because many people in these occupations change jobs frequently, which creates a large number of openings. Keen competition is expected, however, for jobs in popular restaurants and fine dining establishments, where potential earnings from tips are greatest.

Other Customer Services

Customer satisfaction Studies have shown that te overall satisfaction of the guest in a particular hotel determines the likelihood of the client returning to same. The study identifies that the likely influences of the guest or customer satisfaction being room quality and comfort, staff services, security and amenities topping the priority list. Customer satisfaction is very important for a hotel to survive. It is important to give prompt attention to the needs of the customer without delay. This could where a reservation or booking is concerned or even a room in a particular floor and facilities required, giving proper feed back and seeing that it is done at the time the guest checks in is very important.

It could be very annoying for the guest to find out that his/her needs have not been met, and the guest could be put off about the hotel even before he enters. As such, delivering what you have promised, will gain the confidence of the customer.

An important factor in the hospitality industry, but sometimes goes unheeded is being friendly and approachable. The customer should be made comfortable, and they should be treated as if he is the most important guest in the hotel. A friendly and courteous approach will make the customer feel welcomed. As such at all times remaining polite is very important in the hospitality industry.

Maintaining customer service policies are priority. Do not treat one guest different to the other. In case, there are issues have clear rules, where the guest is not sent from pillar to post to sort out issue. Intervene, take the matter upon yourself and resolve so that the customer is aware that no matter what, the hotel will find a solution.

Customers feel important and recognized, when the hotel goes out that extra mile to attend to dtail. Sending out personalized letter, rather than the normal standard hotel leaflets is a sure way to satisfy the customer. Make note of birthdays and anniversaries and other important events during their stay with you. If you had paid attention to detail, you would have these facts with you, or even at the time the reservation is made, the guest would have mentioned it to you. Complimentary notes, flowers or chocolates to the guest could ensure satisfaction. Anticipate your clients need, let it be a book, software, CD or even a requirement, making note of same and supplying it free of charge to the guest, will make the guest feel happy and content.

Some of these will be time consuming but cost effective, some will require just commitment. The existence of the hotel depends on the staff and the management and their approach to the needs and the wants of the guest. As such, taking note and making sure that the guest is satisfied at all cost, will in return bring success and revenue to the hotel.

Make note that one satisfied customer is adequate to promote the hotel to another.

Cashiers

Supermarkets, department stores, gasoline service stations, movie theaters, restaurants, and many other businesses employ cashiers to register the sale of their goods and services. Although specific job duties vary by employer, cashiers usually are assigned to a register at the beginning of their shifts and are given a drawer containing a specific amount of money with which to start—their "till." They must count their till to ensure that it contains the correct amount of money and adequate supplies of change.

Some cashiers also handle returns and exchanges. When they do, they must ensure that returned merchandise is in good condition, and determine where and when it was purchased and what type of payment was used.

After entering charges for all items and subtracting the value of any coupons or special discounts, cashiers total the customer's bill and take payment. Forms of payment include cash, personal checks, and gift, credit, and debit cards. Cashiers must know the store's policies and procedures for each type of payment the store accepts.

For checks and credit and debit card charges, they may request additional identification from the customer or call in for an authorization. They must verify the age of customers purchasing alcohol or tobacco. When the sale is complete, cashiers issue a receipt to the customer and return the appropriate change. They may also wrap or bag the purchase.

At the end of their shifts, cashiers once again count the drawers' contents and compare the totals with sales data. An occasional shortage of small amounts may be overlooked but, in many establishments, repeated shortages are grounds for dismissal. In addition to counting the contents of their drawers at the end of their shifts, cashiers usually separate and total charge forms, return slips, coupons, and any other noncash items.

Most cashiers use scanners and computers, but some establishments still require price and product information to be entered manually. In a store with scanners, a cashier passes a product's Universal Product Code over the scanning device, which transmits the code number to a computer. The computer identifies the item and its price. In other establishments, cashiers manually enter codes into computers and then descriptions of the items and their prices appear on the screen.

Depending on the type of establishment, cashiers may have other duties as well. In many supermarkets, for example, cashiers weigh produce and bulk food, as well as return unwanted items to the shelves. In convenience stores, cashiers may be required to know how to use a variety of machines other than cash registers, and how to furnish money orders and sell lottery tickets. Operating ticket-dispensing machines and answering customers' questions are common duties for cashiers who work at movie theaters and ticket agencies.

Work environment. Most cashiers work indoors, usually standing in booths or behind counters. Often, they are not allowed to leave their workstations without supervisory approval because they are responsible for large sums of money. The work of cashiers can be very repetitious, but improvements in workstation design in many stores are alleviating problems caused by repetitive motion. In addition, the work can sometimes be dangerous; the risk from robberies and homicides is much higher for cashiers than for other workers, although more safety precautions are being taken to help deter robbers.

About 47 percent of all cashiers worked part time in 2008. Hours of work often vary depending on the needs of the employer. Generally, cashiers are expected to work weekends, evenings, and holidays to accommodate customers' needs. However, many employers offer flexible schedules. Because the holiday season is the busiest time for most retailers, many employers restrict the use of vacation time from Thanksgiving through the beginning of January.

Training, other Qualifications, and Advancement

Cashier jobs usually are entry-level positions requiring little or no previous work experience. They require good customer service skills. Education and training. Although there are no specific educational requirements, employers filling full-time jobs often prefer applicants with high school diplomas.

Nearly all cashiers are trained on the job. In small businesses, an experienced worker often trains beginners. The trainee spends the first day observing the operation and becoming familiar with the store's equipment, policies, and procedures. After this, trainees are assigned to a register—frequently under the supervision of an experienced worker. In larger businesses, trainees spend several days in classes before being placed at cash registers. Topics typically covered in class include a description of the industry and the company, store policies and procedures, equipment operation, and security.

Training for experienced workers is not common, except when new equipment is introduced or when procedures change. In these cases, the employer or a representative of the equipment manufacturer trains workers on the job. Other qualifications. People who want to become cashiers should be able to do repetitious work accurately. They also need basic mathematics skills and good manual dexterity. Because cashiers deal constantly with the public, they should be neat in appearance and able to deal tactfully and pleasantly with customers. In addition, some businesses prefer to hire workers who can operate specialized equipment or who have business experience, such as typing, selling, or handling money.

Advancement. Advancement opportunities for cashiers vary. For those working part time, promotion may be to a full-time position. Others advance to head cashier or cash-office clerk. In addition, this job offers a good opportunity to learn about an employer's business and can serve as a steppingstone to a more responsible position.

Employment

Cashiers held about 3.55 million jobs in 2008. Although cashiers are employed in almost every industry, 24 percent of all jobs were in grocery stores. Gasoline stations, department stores, and other retail establishments also employed large numbers of these workers. Outside of retail establishments, many cashiers worked in food services and drinking places.

Job Outlook

Cashiers are expected to grow more slowly than the average for all occupations. Opportunities for full-time and part-time jobs are expected to be good because of the need to replace the large number of workers who leave this occupation.

Employment change. Employment of cashiers is expected to grow by 4 percent between 2008 and 2018 which is slower than the average for all occupations. Continued growth in retail sales is expected, but the rising popularity of purchasing goods online will limit the employment growth of cashiers, although many customers still prefer the traditional method of purchasing goods at stores. Also, the growing use of self-service checkout systems in retail trade, especially at grocery stores, should have an adverse effect on employment of cashiers. These self-checkout systems may outnumber checkouts with cashiers in the future in many establishments.

The impact on job growth for cashiers will largely depend on the public's acceptance of this self-service technology. Job prospects. Opportunities for full-time and part-time cashier jobs should continue to be good because of the need to replace the large number of workers who transfer to other occupations or leave the labour force. There is substantial movement into and out of the occupation because education and training requirements are minimal and the predominance of part-time jobs is attractive to people seeking a short-term source of income rather than a full-time career. Historically, workers under the age of 25 have filled many of the openings in this occupation. In 2008, about 47 percent of all cashiers were 24 years of age or younger.

Because cashiers are needed in businesses and organizations of all types and sizes, job opportunities are found throughout the country. However, job opportunities may vary from year to year because the strength of the economy affects demand for cashiers. Companies tend to hire more cashiers when the economy is strong. Seasonal demand for cashiers also causes fluctuations in employment.

Chefs, Head Cooks, andFood Preparation and Serving Supervisors

Chefs, head cooks, andfood preparation and serving supervisors oversee the daily food service operation of a restaurant or other food service establishment. *Chefs and head cooks* are usually responsible for directing cooks in the kitchen, dealing with food-related concerns, and providing leadership. They are also the most skilled cooks in the kitchen and use their creativity and knowledge of food to develop and prepare recipes. *Food preparation and serving supervisors* oversee the kitchen and non-kitchen staff in a restaurant or food service facility. They may also overseefood preparation workers in fast food, cafeteria, or casual dining restaurants, whre the menu is fairly standard from day to day, or in more formal restaurants, where a chef provides specific guidelines and exacting standards on how to prepare each item.

All of these workers—chefs, head cooks, and food preparation and serving supervisors—hire, train, and supervise staff, prepare cost estimates for food and supplies, set work schedules, order supplies, and ensure that the food service establishment runs efficiently and profitably. Additionally, these workers ensure that sanitation and safety standards are observed and comply with local regulations. Fresh food must be stored and cooked properly, work surfaces and dishes clean and sanitary, and staff and customers safe from illness or injury to avoid being closed by the health department or law enforcement.

While all chefs have a role in preparing the food, developing recipes, determining serving sizes, planning menus, ordering food supplies, and overseeing kitchen operations to ensure uniform quality and presentation of meals, different types of chefs may have unique roles to perform or specialize in certain aspects of the job. *Executive chefs, head cooks, and chefs de cuisine,* are primarily responsible for coordinating the work of the cooks and directing the preparation of meals. Executive chefs are in charge of all food service operations and also may supervise several kitchens of a hotel, restaurant or corporate dining operation. A *sous chef*, or sub chef, is the second-in-command and runs the kitchen in the absence of the chef. Many chefs earn fame both for themselves and for their kitchens because of the quality and distinctive nature of the food they serve.

As a greater variety of establishments prepare and serve food, chefs and head cooks and first-line supervisors of food preparation and serving workers can be found in a greater variety of places. Grocery and specialty food stores employ these workers to develop recipes and prepare meals for customers to carry out. They increasingly work in residential care facilities, such as nursing homes, and in schools and hospitals.

Some chefs and head cooks work for individuals rather than for restaurants, cafeterias, or food manufacturers. *Personal chefs* and *private household cooks* plan and prepare meals in private homes according to the client's tastes or dietary needs. They order groceries and supplies, clean the kitchen, and wash dishes and utensils. They also may serve meals. Personal chefs usually prepare a week's worth of meals in the client's home for the client to heat and serve according to directions. They may be self-employed or work as part of a team of personal chefs and employed by a company that provides this service. Private household cooks typically work full time for one client, such as corporate executives, university presidents, or diplomats, who regularly entertain as a part of their official duties.

While the work of chefs and head cooks is concentrated in the kitchen or in providing overall guidance, food preparation and serving supervisors oversee specific areas of operation in food service establishments or the kitchen and counter areas of quick service restaurants. In fast food and casual dining restaurants, they may share many of the same functions with food service managers. They are responsible for dealing with customer complaints, balancing the books at the end of the day, scheduling workers, and ordering supplies. They also supervise and train kitchen and food preparation staff and ensure that these workers know how to gather food supplies, operate equipment, and assemble orders.

Work environment. Restaurants and other food service facilities where these workers are employed are required to be clean and sanitary. Although the seating areas of eating places are often attractive, kitchens can be crowded and hot and filled with potential dangers, such as hot ovens and slippery floors. Job hazards for those working in kitchens include slips and falls, cuts, and burns, but these injuries are seldom serious. Chefs, head cooks, and supervisors are under constant pressure to get meals prepared quickly, while ensuring quality is maintained and safety and sanitation guidelines are observed. Because the pace can be hectic during peak dining times, workers must be able to communicate clearly so that food orders and service are done correctly.

Work hours in restaurants may include early mornings, late evenings, holidays, and weekends. Schedules for those working in offices, factories and school cafeterias may be more regular. In fine-dining restaurants, work schedules tend to be longer because of the time required to prepare ingredients in advance. Many executive chefs regularly work 12-hour days because they oversee the delivery of foodstuffs early in the day, plan the menu, and prepare those menu items that require the most skill. Depending upon the days of operation, some chefs or other supervisors may take less busy days off to offset the longer hours on other days.

Training, Other Qualifications, and Advancement

Most workers in these occupations have prior experience in the food service or hospitality industries. Most start as food preparation workers or line cooks in a full-service restaurant and work their way up to positions with more responsibility. Some attend cooking school or take vocational training classes and participate in internships or apprenticeship programs to acquire the additional skills needed to create menus and run a business.

Education and training. While most chefs, head cooks, and food preparation and serving supervisors have some postsecondary training, many experienced workers with less education can still be promoted. Formal training may take place at a community college, technical school, culinary arts school, or a 2-year or 4-year college with a degree in hospitality. A growing number of chefs participate in training programs sponsored by independent cooking schools, professional culinary institutes, 2-year or 4-year colleges with a hospitality or culinary arts department, or in the armed forces. Some large hotels and restaurants also operate their own training and job-placement programs for chefs and head cooks. Executive chefs, head cooks, and sous chefs who work in fine-dining restaurants require many years of training and experience.

For students in culinary training programs, most of their time is spent in kitchens learning to prepare meals by practicing cooking skills. They learn knife techniques and proper use and care of kitchen equipment. Training programs also include courses in nutrition, menu planning, portion control, purchasing and inventory methods, proper food storage procedures, and use of leftover food to minimize waste. Students also learn sanitation and public health rules for handling food. Training in food service management, computer accounting and inventory software, and banquet service are featured in some training programs. Most formal training programs also require students to get experience in a commercial kitchen through an internship, apprenticeship, or out-placement program.

Although formal training is an important way to enter the profession, many chefs are trained on the job, receiving real work experience and training from chef-mentors in the restaurants where they work. Others enter the profession through formal apprenticeship programs sponsored by professional culinary institutes, industry associations, and trade unions in coordination with the U.S. Department of Labour. The American Culinary Federation accredits more than 200 formal academic training programs and sponsors apprenticeship programs around the country. Typical apprenticeships last 2 years and combine classroom training and work experience. Accreditation is an indication that a culinary program meets recognized standards regarding course content, facilities, and quality of instruction.

Other qualifications. Chefs, head cooks, and food preparation and serving supervisors must demonstrate strong leadership and communication skills and have the ability to motivate others. Chefs and head cooks also must have an intense desire to cook, be creative, and have a keen sense of taste and smell. Personal cleanliness is essential because most States require health certificates indicating that workers are free from communicable diseases. Knowledge of a foreign language can be an asset because it may improve communication with other restaurant staff, vendors, and the restaurant's clientele.

Certification and advancement. The American Culinary Federation certifies pastry professionals, personal chefs, and culinary educators in addition to various levels of chefs. Certification standards are based primarily on experience and formal training. Although certification is not required, it can help to prove accomplishment and lead to advancement and higher-paying positions. Advancement opportunities for chefs, head cooks, and food preparation and serving supervisors depend on their training, work experience, ability to perform more responsible and sophisticated tasks, and their leadership abilities.

Food preparation and serving supervisors may advance to become food service managers while some chefs and head cooks may go into business as caterers or personal chefs or open their own restaurant. Others may become instructors in culinary training programs, consultants on kitchen design, or food product or equipment sales representatives. A number of chefs and head cooks advance to executive chef positions or food service management positions. When staying in the restaurant business, advancement usually involves moving to a better, busier, or bigger restaurant or working at the corporate level overseeing several restaurants or food service facilities or testing new recipe, menu, or design concepts.

Employment

Chefs, head cooks, and food preparation and serving supervisors held 941,600 jobs in 2008. Food preparation and serving supervisors held 88 percent of these jobs and chefs and head cooks held the remaining 12 percent. Nearly half of chefs and head cooks were employed at full-service restaurants. About nine percent each were employed by hotels and the special food services industry that includes caterers and food service contractors. Eight percent were self-employed.

Forty-three percent of food preparation and serving supervisors were employed by limited-service eating places, made up mostly of cafeterias and fast food restaurants and other places that offer simple carry-out food items. Another 25 percent were employed by full-service restaurants. Supervisors are also found in schools, the special food services industry, and a wide variety of other places that serve food.

Job Outlook

Job opportunities are expected to be good, despite slower than average employment growth, due to the large numbers of workers who leave the occupation and need to be replaced. However, keencompetition is expected for jobs at upscale restaurants that generally pay more.

Employment change. Employment of chefs, head cooks, and food preparation and serving supervisor is expected to increase by 6 percent over the 2008-18 decade, which is more slowly than the average for all occupations.

Growth will be generated by increases in population, a growing variety of dining venues, and continued demand for convenience. As more people opt for the time-saving ease of letting others do the cooking, the need for workers to oversee food preparation and serving will increase. Also, there is a growing consumer desire for healthier, made-from-scratch meals that chefs and head cooks can better prepare.

Job prospects. Job openings for chefs, head cooks, and food preparation and serving supervisors are expected to be good through 2018; however, competition should be keen for jobs at the more upscale restaurants tat tend to pay more. Workers with a good business sense will have beter job prospects, especially at restaurant chains where attention to costs is very important. Although job growth will create many new positions, the majority of job openings will stem from the need to replace workers who leave the occupation. The fast pace, long hours, and high energy levels required for these jobs often lead to high turnover.

Flight attendants

- Competition for positions is expeted to remain keen because the opportunity for travel attracts more applicants than there are jobs.
- Job duties are learned through formal on-the-job training at a flight training centre.
- A high school diploma is the minimum educational requirement, but airlines prefer applicants with a college degree and with experience in dealing with the public.

Nature of the Work

Major airlines are required by law to provide *flight attendants* for the safety and security of the traveling public. Although the primary job of the flight attendants is to ensure that security and safety regulations are followed, attendants also try to make flights comfortable and enjoyable for passenger.

At least 1 hour before takeoff, attendants are biefed by the captain—the pilot in command—on such things as emergency evacuation procedures, coordination of the crew, the length of the flight, expected weather conditions, and any special issues having to do with passengers. Flight attendants make sure that first-aid kits and other emergency equipment are aboard and in working order and that the passenger cabin is in order, with adequate supplesof food, beverages, and any other amenities. As passengers board the plane, flight attendants greet them, check their tickets, and tell them where to store carry-on items.

Before the plane takes off, flight attendants instructall passengers in the use of emergency equipment and check to see that seatbelts are fastened, seat backs are in upright positions, and all carry-on items are properly stowed. In the air, helping passengers in the event of an emergency is the most important responsibility of a flight attendant. Safety-related actions range from reassuring passengers during rough weather to directing passengers who must evacuate a plane following an emergency landing.

Flight attendants also answer questions about the flight, and help small children, elderly or disabled persons, and any others needing assistance. Flight attendants may administer first aid to passengers who become ill. Flight attendants generally serve beverages and on many flights sell precooked meals or snacks. Prior to landing, flight attendants take inventory of headsets, alcoholic beverages, and moneys collected. They also report any medical problems passengers may have had, the condition of cabin equipment, and any lost-and-found articles.

Lead, or first, flight attendants, sometimes known as *pursers*, oversee the work of the other attendants aboard the aircraft, while performing most of the same duties.

Work environment. Because airlines operate around the clock and year round, flight attendants can work nights, holidays, and weekends. In most cases, agreements between the airline and the employees' union determine the total daily and monthly working time. Scheduled on-duty time usually is limited to 12 hours per day, however flight attendants can be scheduled up to 14 hours per day, with somewhat greater maximums for international flying. The Federal Aviation Administration (FAA) requires that flight attendants receive 9 consecutive hours of rest following any duty period.

Attendants usually fly 65 to 90 hours a month and generally spend another 50 hours a month on the ground preparing planes for flights, writing reports following completed flights, and waiting for planes to arrive. Most airlines guarantee a minimum of 65 to 85 flight hours per month, with the option to work additional hours. Flight attendants receive extra compensation for additional hours.

Flight attendants may be away from their home base at least one-third of the time. During this period, the airlines provide hotel accommodations and an allowance for meal expenses.

Flight attendants must be flexible and willing to relocate. However, many flight attendants elect to live in one place and commute to their assigned home base. Home bases and routes worked are bid for and awarded on a seniority basis, so the longer the flight attendant has been employed, the more likely he or she is to work on their preferred flights. Almost all flight attendants start out working on reserve status, or on call. Flight attendants on reserve status usually live near their home base, because they are required to be able to report to their home base on short notice. On small corporate airlines, flight attendants often work on an as-needed basis and must adapt to varying environments and passengers.

The combination of free time and free or discounted airfares provides flight attendants the opportunity to travel. However, the work can be strenuous and trying. Flight attendants stand during much of the flight and must remain pleasant and efficient, regardless of how tired they are or how demanding passengers may be. Occasionally, flight attendants must deal with turbulent flights which can cause difficulties regarding service and cause anxiety among passengers that flight attendants must address.

Working in a moving aircraft leaves flight attendants susceptible to injuries. According to BLS data, full-time flight attendants experienced a much higher than average work-related injury and illness rate. Various physical injuries can occur when opening overhead compartments or while pushing heavy service carts. In addition, medical problems can arise from irregular sleeping and eating patterns, dealing with stressful passengers, working in a pressurized environment, and breathing recycled air.

Training, Other Qualifications, and Advancement

Flight attendants must be certified by the FAA. A high school diploma or its equivalent is the minimum educational requirement, but airlines increasingly prefer applicants who have a college degree. Experience in dealing with the public is important, because flight attendants must be able to interact comfortably with strangers and remain calm under duress.

Education and training. A high school diploma or its equivalent is the minimum educational requirement. However, airlines increasingly prefer applicants with a college degree. Applicants who attend schools or colleges that offer flight attendant training may have an advantage over other applicants. Highly desirable areas of concentration include people-oriented disciplines, such as communications, psychology, nursing, travel and tourism, hospitality, and education. Flight attendants for international airlines generally must speak a foreign language fluently. For their international flights, some of the major airlines prefer candidates who can speak two major foreign languages.

Once hired, all candidates must undergo a period of formal training. The length of training, ranging from 3 to 6 weeks, depends on the size and type of carrier and takes place at the airline's flight training centre. Airlines that do not operate training centres generally send new employees to the centre of another airline. Some airlines may provide transportation to the training centres and an allowance for room, board, and school supplies, while other airlines charge individuals for training. New trainees are not considered employees of the airline until they successfully complete the training program. Trainees learn emergency procedures, such as evacuating an airplane, operating emergency systems and equipment, administering first aid, and surviving in the water. In addition, trainees are taught how to deal with disruptive passengers and with hijacking and terrorist situations.

New hires learn flight regulations and duties, gain knowledge of company operations and policies, and receive instruction on personal grooming and weight control. Trainees for international routes get additional instruction in passport and customs regulations. Trainees must perform many drills and duties unaided, in front of the training staff. Throughout training, they also take tests designed to eliminate unsuccessful trainees. Toward the end of their training, students go on practice flights. Upon successful completion of training, flight attendants receive the FAA Certificate of Demonstrated Proficiency. Flight attendants also are required to go through periodic retraining and pass an FAA safety examination to continue flying.

Licensure and certification. All flight attendants must be certified by the FAA. To be certified, flight attendants are required to successfully complete training requirements, such as evacuation, fire fighting, medical emergency, and security procedures established by the FAA and the Transportation Security Administration. They also must perform the assigned duties of a cabin crew member and complete an approved proficiency check. Flight attendants are certified for specific types of aircraft, regardless of the carrier. Therefore, only 1-day or 2-day recurrent training, with the new carrier, is needed for those flight attendants who change airlines, as long as the type of aircraft remains the same.

Other qualifications. Airlines prefer to hire poised, tactful, and resourceful people who can speak clearly and interact comfortably with strangers and remain calm under duress. Applicants with previous experience in dealing with the public are preferred by airlines. Additionally, airlines usually have age, physical, and appearance requirements. Applicants usually must be at least 18 to 21 years old, although some carriers may have higher minimum-age requirements. Applicants must meet height requirements for reaching overhead bins, which often contain emergency equipment, and most airlines want candidates with weight proportionate to height.

Flight attendants must be in excellent health, and a medical evaluation is required. Vision is required to be correctable to 20/30 or better with glasses or contact lenses (uncorrected no worse than 20/200). Men must have their hair cut above the collar and be clean shaven. Airlines prefer applicants with no visible tattoos, body piercing, or unusual hairstyles or makeup.

In addition to education and training, airlines conduct a thorough background check, which goes back as many as 10 years, as required by the FAA,. Everything about an applicant is investigated, including date of birth, employment history, criminal record, school records, and any gaps in employment. Employment is contingent on a successful background check. An applicant will not be offered a job or will be immediately dismissed if his or her background check shows any discrepancies. All U.S. airlines require that applicants be citizens of the United States or registered aliens with legal rights to obtain employment in the United States.

Advancement. After completing initial training, flight attendants are assigned to one of their airline's bases. New flight attendants are placed on reserve status and are called either to staff extra flights or to fill in for crewmembers that are sick, on vacation, or rerouted. When they are not on duty, reserve flight attendants must be available to report for flights on short notice. They usually remain on reserve for at least 1 year but, in some cities, it may take 5 to 10 years—or longer—to advance from reserve status.

Flight attendants who no longer are on reserve bid monthly for regular assignments. Because assignments are based on seniority, usually only the most experienced attendants get their choice of assignments. Advancement takes longer today than in the past, because experienced flight attendants are remaining in this career longer than in the past.

Some flight attendants become supervisors, moving from senior or lead flight attendant, to check flight attendant, to flight attendant supervisor, then on to base manager, and finally to manager or vice president of in-flight operations. They may take on additional duties, such as recruiting, instructing, or developing in-flight products. Their experience also may qualify them for numerous airline-related jobs involving contact with the public, such as reservation ticket agent or public relations specialist. Flight attendants who do not want to travel often for various reasons may move to a position as an administrative assistant. With additional education, some flight attendants may decide to transfer to other areas of the airline for which they work, such as risk management or human resources.

Employment

Flight attendants held about 98,700 jobs in 2008. Commercial airlines employed the vast majority of flight attendants, and most attendants lived near major metropolitan airports or airports operating as hubs for the major airlines. A small number of flight attendants worked for companies that offered chartered flights.

Job Outlook

Employment of flight attendants is projected to grow about as fast as average. Competition for jobs is expected to remain keen because the opportunity for travel attracts more applicants than there are jobs.

Employment change. Employment of flight attendants is expected to grow by 8 percent, which is about as fast as the average for all occupations over the 2008–18 period. Population growth and an improving economy are expected to boost the number of airline passengers. As airlines expand their capacity to meet rising demand by increasing the number and size of planes in operation and the number of flights offered, more flight attendants will be needed.

Job prospects. Despite growing demand for flight attendants, competition is expected to be keen because this job usually attracts more applicants than there are jobs, with only the most qualified eventually being hired. College graduates who have experience dealing with the public should have the best chance of being hired. Job opportunities may be better with the faster growing regional and commuter, low-cost, and charter airlines. There also are job opportunities for professionally trained flight attendants to work for companies operating private aircraft for their executives.

The majority of job opportunities through the year 2018 will arise from the need to replace flight attendants who leave the labour force or transfer to other occupations, often for higher earnings or a more stable lifestyle. With the job now viewed increasingly as a profession, however, fewer flight attendants leave their jobs, and job turnover is not as high as in the past. According to the Association of Flight Attendants, the average job tenure of attendants is currently 16 years and is increasing.

In the long run, opportunities for persons seeking flight attendant jobs should improve as the airline industry expands. Over the next decade, however, demand for flight attendants will fluctuate with the demand for air travel, which is highly sensitive to swings in the economy. During downturns, as air traffic declines, the hiring of flight attendants declines, and some experienced attendants may be laid off until traffic recovers.

Earnings

Median annual wages of flight attendants were $35,930 in May 2008. The middle 50 percent earned between $28,420 and $49,910. The lowest 10 percent earned less than $20,580, and the highest 10 percent earned more than $65,350. According to data from the Association of Flight Attendants, beginning attendants had median earnings of $16,191 a year in 2009. Beginning pay scales for flight attendants vary by carrier, however. New hires usually begin at the same pay scale, regardless of experience; all flight attendants receive the same future pay increases based on an established pay scale.

Some airlines offer incentive pay for working holidays, night and international flights, or taking positions that require additional responsibility or paperwork.

Flight attendants and their immediate families are entitled to free or discounted fares on their own airline and reduced fares on most other airlines. Some airlines require that the flight attendant be with an airline for 3 to 6 months before taking advantage of this benefit. Other benefits may include medical, dental, and life insurance; 401K or other retirement plan; sick leave; paid holidays; stock options; paid vacations; and tuition reimbursement. Flight attendants also receive a "per diem" allowance for meal expenses while on duty away from home. Flight attendants are required to purchase uniforms and wear them while on duty. The airlines usually pay for uniform replacement items, and may provide a small allowance to cover cleaning and upkeep of the uniforms.

The majority of flight attendants hold union membership, primarily with the Association of Flight Attendants. Other unions that represent flight attendants include the Transport Workers Union of America and the International Brotherhood of Teamsters.

Retail Salespersons

- Good employment opportunities are expected because of the need to replace the large number of workers who leave the occupation each year.
- Many salespersons work evenings and weekends, particularly during peak retail periods.
- Employers look for people who enjoy working with others and who have good communication skills, an interest in sales work, a neat appearance, and a courteous demeanor.
- Although advancement opportunities are limited, having a college degree or a great deal of experience may help retail salespersons move into management positions.

Nature of the Work

Whether selling shoes, computer equipment, or automobiles, retail salespersons assist customers in finding what they are looking for. They also try to increase sales by describing a product's features, demonstrating its uses, and promoting its value.

In addition to selling, many retail salespersons—especially those who work in department and apparel stores—conduct financial transactions with their customers. This usually involves receiving payments by cash, check, debit card, or credit card; operating cash registers; and bagging or packaging purchases. Depending on the hours they work, retail salespersons may have to open or close cash registers. This work may include counting the money in the register and separating charge slips, coupons, and exchange vouchers.

Retail salespersons also may have to make deposits at a cash office. (Cashiers, who have similar duties, are discussed elsewhere in the *Handbook*.) In addition, retail salespersons may help stock shelves or racks, arrange for mailing or delivery of purchases, mark price tags, take inventory, and prepare displays.

For some sales jobs, particularly those involving expensive and complex items, retail salespersons need special knowledge or skills.

For example, salespersons who sell automobiles must be able to explain the features of various models, the manufacturers' specifications, the types of options and financing available, and the details of associated warranties. In addition, all retail salespersons must recognize security risks and thefts and understand their organization's procedure for handling such situations—procedures that may include notifying security guards or calling police. Work environment. Most retail salespersons work in clean, comfortable, well-lit stores. However, they often stand for long periods and may need supervisory approval to leave the sales floor. They also may work outdoors if they sell items such as cars, plants, or lumber yard materials.

The Monday-through-Friday, 9-to-5 workweek is the exception rather than the rule for retail salespersons. Many salespersons work evenings and weekends, particularly during holidays and other peak sales periods. The end-of-year holiday season often is the busiest time, and as a result, many employers limit the use of vacation time between Thanksgiving and the beginning of January.

This occupation offers opportunities for both full-time and part-time work. About 34 percent of retail salespersons worked part time in 2008. Part-time opportunities may vary by setting, however, as many who sell big-ticket items are required to work full time.

Training, Other Qualifications, and Advancement

Retail salespersons typically learn their skills through on-the-job training. Although advancement opportunities are limited, having a college degree or a great deal of experience may help retail salespersons move into management positions. Education and training. There usually are no formal education requirements for retail sales positions, but employers often prefer applicants with a high school diploma or its equivalent. This may be especially important for those who sell technical products or "big-ticket" items, such as electronics or automobiles. A college degree may be required for management trainee positions, especially in larger retail establishments.

Most retail salespersons receive on-the job training, which usually lasts anywhere from a few days to a few months. In small stores, newly hired workers usually are trained by an experienced employee. In large stores, training programs are more formal and generally are conducted over several days. Topics often include customer service, security, the store's policies and procedures, and cash register operation. Depending on the type of product they are selling, employees may be given additional specialized training. For example, those working in cosmetics receive instruction on the types of products the store offers and for whom the cosmetics would be most beneficial. Likewise, those who sell computers may be instructed in the technical differences between computer products. Because providing the best possible service to customers is a high priority for many employers, employees often are given periodic training to update and refine their skills.

Other qualifications. Employers look for people who enjoy working with others and who possess good communication skills. Employers also value workers who have the tact and patience to deal with difficult customers. Among other desirable characteristics are an interest in sales work, a neat appearance, and a courteous demeanor. The ability to speak more than one language may be helpful for employment in communities where people from various cultures live and shop. Before hiring a salesperson, some employers conduct a background check, especially for a job selling high-priced items.

Advancement. Opportunities for advancement vary. In some small establishments, advancement is limited because one person—often the owner—does most of the managerial work. In others, some salespersons can be promoted to assistant manager. Large retail businesses usually prefer to hire college graduates as management trainees, making a college education increasingly important. However, motivated and capable employees without college degrees still may advance to administrative or supervisory positions in large establishments.

As salespersons gain experience and seniority, they often move into positions with greater responsibility and may be given their choice of departments in which to work. This opportunity often means moving to areas with higher potential earnings and commissions. The highest earnings potential usually lies in selling "big-ticket" items—such as cars, jewellery, furniture, and electronic equipment—although doing so often requires extensive knowledge of the product and an excellent talent for persuasion.

Previous sales experience may be an asset when one is applying for positions with larger retailers or in nonretail industries, such as financial services, wholesale trade, or manufacturing.

Employment

Retail salespersons held about 4.5 million jobs in 2008. The largest employers were clothing and clothing accessories stores, department stores, building material and supplies dealers, motor vehicle and parts dealers, and general merchandise stores such as warehouse clubs and supercenters. In addition, about 156,500 retail salespersons were self-employed. Because retail stores are found in every city and town, employment is distributed geographically in much the same way as the population.

Job Outlook

Employment is expected to grow about as fast as average. Due to the frequency with which people leave this occupation, job opportunities are expected to be good. Employment change. Employment is expected to grow by 8 percent over the 2008–18 decade, about as fast as the average for all occupations. In addition, given the size of this occupation, about 374,700 new retail salesperson jobs will arise over the projections decade—more jobs than will be generated in almost any other occupation.

Employment growth among retail salespersons reflects rising retail sales stemming from a growing population. Many retail establishments will continue to expand in size and number, leading to new retail sales positions. Growth will be fastest in general merchandise stores, many of which sell a wide assortment of goods at low prices. As consumers continue to prefer these stores other establishments with higher prices, growth in this industry will be rapid. Employment of retail sales persons is expected to decline in department stores and automobile dealers as these industries see a reduction in store locations.

Despite the growing popularity of electronic commerce, the impact of online shopping on the employment of retail salespersons is expected to be minimal. Internet sales have not decreased the need for retail salespersons. Retail stores commonly use an online presence to complement their in-store sales, and many consumers prefer to buy merchandise in person. Retail salespersons will remain important in assisting customers, providing specialized service, and increasing customer satisfaction.

Job prospects. Employment opportunities for retail salespersons are expected to be good because of the need to replace the large number of workers who transfer to other occupations or leave the labour force each year. In addition, many new jobs will be created for retail salespersons as businesses seek to expand operations and enhance customer service. A substantial number of these openings should occur in warehouse clubs and supercenters as a result of strong growth among these establishments.

Opportunities for part-time work should be abundant, and demand is expected be strong for temporary workers during peak selling periods, such as the end-of-year holiday season between Thanksgiving and the beginning of January. During economic downturns, sales volumes and the resulting demand for sales workers usually decline.

Consequently, retail sales jobs generally are more susceptible to fluctuations in the economy than are many other occupations.

Earnings

Median hourly wages of wage-and-salary retail salespersons, including commissions, were $9.86 in May 2008. The middle 50 percent earned between $8.26 and $13.35 an hour. The lowest 10 percent earned less than $7.37, and the highest 10 percent earned more than $19.14 an hour. Median hourly wages in the industries employing the largest numbers of retail salespersons in May 2008 were as follows:

Automobile dealers	$18.91
Building material and supplies dealers	11.95
Other general merchandise stores	9.22
Department stores	9.14
Clothing stores	8.94

Many beginning or inexperienced workers earn the Federal minimum wage of $7.25 an hour, but many States set minimum wages higher than the Federal minimum. In areas where employers have difficulty attracting and retaining workers, wages tend to be higher than the legislated minimum.

Compensation systems can vary by type of establishment and merchandise sold. Salespersons receive hourly wages, commissions, or a combination of the two. Under a commission system, salespersons receive a percentage of the sales they make. This system offers sales workers the opportunity to increase their earnings considerably, but they may find that their earnings depend strongly on their ability to sell their product and on the ups and downs of the economy.

Benefits may be limited in smaller stores, but benefits in large establishments usually are considerable. In addition, nearly all salespersons are able to buy their store's merchandise at a discount, with the savings depending on the type of merchandise. Also, to bolster revenue, employers may use incentive programs such as awards, bonuses, and profit-sharing plans to the sales staff.

Housekeeping

Housekeeping or housecleaning is the systematic process of making a home neat and clean in approximately that order. This maybe applied more broadly that just an individual home, or as a metaphor for a similar "clean up" process applied elsewhere such as a procedural reform. It can also be called household management, which is the act of overseeing the organizational, financial, day-to-day operations of a house or estate, and the managing of other domestic concerns.

In the process of housekeeping general cleaning activities are completed, such as disposing of rubbish, storing of belongings in regular places, cleaning dirty surfaces, dusting and vacuuming. It is also the care and control of property, ensuring its maintenance and proper use and appearance. In a hotel, "housekeeping" is also a term for the cleaning personnel.

Some housekeeping is housecleaning and some housekeeping is home chores. Home chores are housework that needs to be done at regular intervals, Housekeeping includes the budget and control of expenditures, preparing meals and buying food, paying the heat bill, and cleaning the house. Outdoor housecleaning chores include removing leaves from rain gutters, washing windows, sweeping doormats, cleaning the pool, putting away lawn furniture, and taking out the trash.

Tools include the vacuum cleaner, broom and mop. Supplies such as cleaning solutions and sponges are sold in grocery stores and elsewhere. Professional cleaners can be hired for less frequent or specialist tasks such as cleaning blinds, rugs, and sofas. Professional services are also offered for the basic tasks. Safety is a consideration because some cleaning products are toxic and some cleaning tasks are physically demanding. *Green cleaning* refers to cleaning without causing pollution. The history of housecleaning has links to the advancement of technology.

Reasons

People perform house cleaning for the home to look better and be safer and easier to live in. It is in response to clutter, disorder, litter, dirtiness or to prevent such. Without housecleaning limescale builds up on taps, mold grows in wet areas, bacterial action make the garbage disposal and toilet smell and cobwebs accumulate. With organisation belongings are easily found and table tops are clear.

Indoor Litter

Disposal of rubbish is an important aspect of house cleaning, the reasons for this are psychological, social and practical. Plastic bags are designed and manufactured specifically for the collection of litter. Many are sized to fit common waste baskets and trash cans. Paper bags are made to carry aluminum cans, glass jars and other things. Recycling is possible with some kinds of litter.

Clutter Problem

Clutter is belongings that have not been put away into storage locations designed for them. If there are not enough shelves and drawers and hangers and there can't be more, this is a cause of clutter. There is a limit to the number of possessions that can be neatly stored in a home. A tangled pile of old coats, mittens, scarves, hats and boots occurs in some hall closets. Pawnbrokers, thrift shops and garbage collectors are involved in the prevention of clutter.

Dustiness

Over time dust accumulates on household surfaces. As well as making the surfaces dirty, when dust is disturbed it can become suspended in the air, causing sneezing and breathing trouble. It can also transfer from furniture to clothing, making it unclean. Various tools have been invented for dust removal; Feather and lamb's wool dusters, cotton and polyester dust cloths, disposable paper "dust cloths", dust mops for smooth floors and vacuum cleaners. Vacuum cleaners often have a variety of tools to enable them to remove not just from carpets and rugs, but from hard surfaces and upholstery. This way can help your things well organize. Things will look so clean likewise neat.

Dirtiness

Although one meaning of housecleaning is "improvement," the primary meaning of housecleaning is "the cleaning of a house." A goal of housecleaning would be the removal of mold from shower grout or smudges and splatters from a kitchen wall. Examples of dirtiness or "soil" would be dry coffee spills and jelly drips or muddy footprints on carpet. Soap and water is a "cleaner." Equipment used with a cleaner might be a bucket and sponge. A modern tool is the spray bottle, but the scientific principle is the same. Cleaning supplies have directions and are sold at janitorial and other stores. Cleaning specialists such as carpet cleaners are listed in phone books. Housecleaning is done to achieve and to enjoy a cleaner house.

Training

In the early 1800s throughout the industrialized world there was a campaign to teach girls the domestic arts (cooking, cleaning, hygiene, sewing, art, decor, etc) in school. In the early 1870s there were college courses in home economics and by 1880 there were high school courses. There continues to be high school and vocational school courses and college degrees in home economics, which prepares students for various employment, as well as home and family management. Home economists are taught and teach about relationships, children, economy, shopping, management of home and family, sewing and interior decoration. Although boys have not been required to learn cooking and cleaning in school, a few have taken a course or two.

Household Chemicals

Various household cleaning products have been developed to facilitate the removal of dust and dirt, for surface maintenance, and for disinfection. Products are available in powder, liquid or spray form. The basic ingredients determine the type of cleaning tasks for which they are suitable. Some are packaged as general purpose cleaning materials whilst others are targeted at specific cleaning tasks such as drain clearing, oven cleaning, lime scale removal and polishing furniture. Household cleaning products provide aesthetic and hygiene benefits but are also associated with health risks for the users, and building occupants. The US Department of Health and Human Services offers the public access to the Household Products Database.

This database provides consumer information for over 4,000 products based on information provided by the manufacturer through the Material Safety Data Sheet. Consumers can search for products by brand name, manufacturer, ingredients, product recalls, and health effects. Surfactants lower the surface tension of water, making it able to flow into smaller tiny cracks and crevices in soils making removal easier. Alkaline chemicals break down known soils such as grease and mud. Acids break down soils such as lime scale, soap scum, and stains of mustard, coffee, tea, and alcoholic beverages. Some solvent-based products are flammable and some can dissolve paint and varnish. Disinfectants stop smell and stains caused by bacteria.

When multiple chemicals are applied to the same surface without full removal of the earlier substance, the chemicals may interact. This interaction may result in a reduction of the efficiency of the chemicals applied (such as a change in pH value caused by mixing alkalis and acids) and in cases may even emit toxic fumes. An example of this is the mixing of ammonia-based cleaners (or acid-based cleaners) and bleach. This causes the production of chloramines that volatilize (become gaseous) causing acute inflammation of the lungs (toxic pneumonitis), long-term respiratory damage, and potential death.

Residue from cleaning products and cleaning activity (dusting, vacuuming, sweeping) have been shown to impact indoor air quality (IAQ) by redistributing particulate matter (dust, dirt, human skin cells, organic matter, animal dander, particles from combustion, fibers from insulation, pollen, and polycyclic aromatic hydrocarbons that gaseous or liquid particles become adsorbed to. The particulate matter and chemical residual will of be highest concentrations right after cleaning but will decrease over time depending upon levels of contaminants, air exchange rate, and other sources of chemical residual. Of most concern are the family of chemicals called VOCs such as formaldehyde, toluene, and limonene.

Volatile organic compounds (VOCs) are released from many household cleaning products such as disinfectants, polishes, floor waxes, air-freshening sprays, all purpose cleaning sprays, and glass cleaner. These products have been shown to emit irritating vapors. VOCs are of most concern due to their tendency to evaporate and be inhaled into the lungs or adsorbed to existing dust, which can also be inhaled. It has been found that aerosolized (spray) cleaning products are important risk factors and may aggravate symptoms of adult asthma, respiratory irritation, childhood asthma, wheeze, bronchitis, and allergy.

Other modes of exposure to potentially harmful household cleaning chemicals include absorption through the skin (dermis), accidental ingestion, and accidental splashing into the eyes. Products for the application and safe use of the chemicals are also available, such as nylon scrub sponge and rubber gloves. It is up to the consumer to keep themselves safe while using these chemicals. Reading and comprehending the labels is important.

There is a growing consumer and governmental interest in natural cleaning products and green cleaning methods. The use of nontoxic household chemicals is growing as consumers become more informed of the health effects of many household chemicals, and municipalities are having to deal with the expensive disposal of household hazardous waste (HHW).

Tools

"Modern housecleaning tools" is almost an oxymoron. There are few areas of employment where someone from 50 years ago could step into the same job today, but housecleaning is one area where there has been very little change. Brooms remove debris from floors. Brushes clean cracks and crevices. Buckets hold cleaning and rinsing solutions. Carpet sweepers remove surface dust and debris. Chamois do lint-free drying of windows. Clotheslines hold doormats and rugs for cleaning. Clothespins fasten things on clotheslines. Dishwashers are machines that wash dishes placed inside them. Dryers dry textiles placed inside them. Dusters are soft cloths for wiping dust, or are various dust sweepers. Dustpans carry dust and debris swept into them.

Floor machines remove and apply floor finish. Rubber gloves protect hands from dish water and other cleaners. Dust mops remove dust from smooth floors. Wet mops are for washing floors. Paper towels wipe up grease and other materials difficult to rinse from a cloth. Polishing cloths are for applying polish and removing excess polish. Sponges apply cleaning solution and remove soil. Spray bottles apply cleaning solution to be removed by another tool. Squeegees remove solutions from glass. Steam cleaners are machines that help wash carpet. Vacuum cleaners remove dust from inside carpet. *Upright vacuums* are best at carpet vacuuming. *Canister vacuums* are best at upholstery and bare floors. *Wet-dry vacuums* can be used to remove spills and spot removers. Washing machines clean textiles such as dust cloths.

Yard

A home's yard and exterior are sometimes subject to cleaning. The first impression of a home's cleanliness is given by its yard and entranceway. Exterior cleaning also occurs for safety, upkeep and usefulness. It includes removal of paper litter and grass growing in sidewalk cracks. Rain gutters, doormats, pools and the screens and glass of windows are also cleanable. Yard junk-removal might occur and porch clutter removal. The paint of door frames might be washed or an old pinata thrown away.

Professional Cleaning Services

There are many professional house cleaning services located around the world. In recent years, they have grown in popularity due to the increase in both spouses needing to work.

Chapter 3

Beverage Industry

The food industry is on a roller coaster ride as Indians continue to have a feast. Fuelled by– large disposable incomes – the food sector is witnessing a remarkable change in consumption patterns, especially in terms of food.

Food retailing has come of age — from a period when food items were sold in small road side grocer shops & mandis, haats and bazzars by vendors to a stage when food products (processed and groceries) are retailed through supermarket stores where consumers can inspect, select and pick up the products they like in a comfortable ambience and still pay a fair price for the product and the merchandise and sometimes even pay less than the price they would have paid at the nearest food stores. Shopping for groceries is no longer a strenuous and uncomfortable affair. Instead, it is a pleasurable experience. From simple trading activity, food retailing is now heading to the status of an industry. Imagine yourself walking through the air conditioned lanes smelling fresh food and groceries, enjoying light music, experiencing five star ambiences and above all buying food products (vegetables, spices and beverages) without bargaining with the vendors.

Food retail has surpassed the dominating apparel and accessories sector. "Contrary to the belief that fashion is the largest segment of organised retail in India, food & beverages is the major segment, worth Rs. 8,97,000 crore," said Arvind Singhal, Chairman, Technopak, at the Indian Retail Forum held in Mumbai. There is an old industry saying that " customer is king" Food retailers today will update that saying to "customers is the dictator" because of fierce competition, new technology and business practices the market power of customer is strong and growing stronger.

The Food Reta]il Tale

Traditionally, Indians were used to buying their sugar, wheat, pulses, rice etc. from their neighborhood baniya. The majority of food and food products were and still are retailed through neighborhood kirana (baniya) stores. A typical kirana store has a retail area of 200 sq. ft. and sells 500 to 800 stock keeping units (SKUs).

The kirana stores focus on dry food products because the infrastructure for cold storage is lacking. The majority of fresh produce is sold from the carts of travelling vendors. Such produce is deemed to be of low product quality, variety and hygiene. This concept is still popular in B-class and II Tier cities of India and giving employment to thousands of its inhabitants.

Initially the food retail format was seen in A-class cities like Mumbai, Delhi, Chennai which had co-operative stores like "Apna Bazaar" in Mumbai and "Kendriya Bhandar" in Delhi. Both were very successful and are operating many outlets in all strategic localities in the city.

Escorts group in the late eighties diversified into non-auto sectors by getting into agri business or food business. It came out with first "Nanz" store at South Extension in Delhi in 1990.

The first visible sign of the change in food retailing was seen in mid-eighties. Around that time a few new food stores were set up in all metro cities in India. Calcutta was the only exception where it started a little later. At that time couple of leading food stores started operating such as "Morning Stores" and "Modern Stores" in Delhi, "Nilgiri" in Bangalore, and "Food Land" in Mumbai "Spencers Food Stores" in Chennai.

Spencers were the first to tie up with a Singapore based large retail chain — "Dairyland" and had set up the food stores in Chennai. This was a technology tie up. In Mumbai, Garware group during the late eighties had set up a large food store, which is now reported to have been closed down.

Until the late 1990s, food retailing has been concentrated in the south of the country. Southern India has been witnessing revolutionized activity in food retailing.

It has thoroughly experienced the food retailing in various formats such as the supermarkets, hypermarkets and neighbourhood stores. These include Food World, Subhiksha, Nilgiris, Margin Free, and Big Bazaar.

The reason being that most entrepreneurs who started organised retail came from southern India and the cost of real estate in the southern region was less than other regions (particularly north and west). Since then, however, organised food retailing has emerged across the country, inspired by the presence of high potential markets in the north, west and east as well as the success of some non-food retailers and food services companies in these regions.

Neighbourhood Stores

In India about 90% of food purchases are made within a distance of 1.5 km from the customer's home. This means that an organised retailer would need to have a 'neighbourhood store' close to customers in order to capture the share of wallet that is spent on food.

These stores would cater to the consumer's daily and weekly needs. The outlets closest to a neighbourhood store in India are 'Safal' outlets operated by Mother Dairy in Delhi, Margin Free in Kerala and Subhiksha.

Food Research Institutions in India

Produce Little Value

Food Science and Technology Education in India was initially masterminded by pure scientists in the field of chemistry, applied chemistry and biochemistry. The boderline between the basic science and technology and engineering was therefore not clearly demarcated at the initial stage.

As a result, research projects that were undertaken in most of our premium food research institutions including Central Food Technological Research Institute (CFTRI) were designed on the line of basic science exploration in the areas of biochemistry, protein chemistry, nutrition and food processing. Those who have ventured into the processing and technology also focussed only on physiochemical changes of food during processing and process parameters optimization.

Those who have ventured a little ahead thus incorporated process conditions standardisations to improve product quality. Nothing therefore happened in terms of any major breakthrough in process technology in India.

Our food research institutes including universities did very little in terms of technology upgradation and new product development which are of any commercial significance. Whatever little we have seen in new product development and product quality upgradation is the result of the efforts on the part of the industry largely triggered from the need as a survival strategy in the face of competition.

Some of these research institutions also suffer from the syndrome of reinventing the wheel. Crores of rupees have been spent by young food researchers reproducing the knowledge which are either known and or explored elsewhere. This therefore has put the industry, in particular and the nation, in general into the back gear.

We have to painfully observe in various conferences and seminars that the research findings are presented with lot of gusto which are nothing more than the age old findings and established processes.

Another factor that has marked the failure of food research activities in India to create any dent in terms of value addition in the processing and technology is due to the fact that there was little or no collaborative efforts between industry and research institutions.

The mutual mistrust and lack of confidence on the part of the industry leaders on the ability of the institutional research to deliver value has increased with the passage of time. As a result, researchers have ventured into their own chosen subjects while industries have decided to go it alone.

Formation of an independent Ministry of Food Processing Industry which has now been reduced to the status of a department did not help change this scenario. The result is inevitable. While we can see a sea change in other industry categories in terms of product quality and value addition, in case of processed food industry changes are miniscule to say the least, and that too largely due to the efforts of the industry. On the top of it, some of our premium institutions are in reality in a state of decay in terms of quality of research and education.

The research activities that have been initiated earlier by the pioneers in food research and education have lost the inertia in the later years and thus died down. In this context we can remember the name of Prof. V. Subramaniam, first Director of CFTRI, Prof. A.N. Bose, first Director of Central Marine Product Research Institute in Cochin, Prof. A. Sreenivasan-First Director of Food Technology Dept. of BARC. Prof. Gopalan, first Director of Nutrition Research Institute (NII) at Hyderabad etc. CFTRI has seen some of the stalwarts in later years which include the names like Mr. C.P. Natarajan, Dr. H.A.B. Parpia, Dr. B.L. Amia and few others.

National Dairy Research Institute (NDRI) in my reckoning, do have a creditable record of quality research output and in that also contribution of Dr. K.K. Iya and most importantly that of Dr. N. Ganguly are noteworthy.

In oil technology Dr. K.T. Acharya's and Prof. M.M. Chakraborti's contribution is also significant. In the areas of biotechnology and biochemical engineering Prof. T.K. Ghose's contribution is pioneering.

He has, in fact, given the shape to biochemical engineering education and research in India and was associated with Jadavpur University, HBTI-Kanpur and IIT-Delhi. Defece Food Research Laboratory (DFRL) in Mysore also worked on projects of specific interest to our defence forces.

In the areas of research and education we now have a generation gap. After the retirement and or passing away of some of these above mentioned stalwarts most of these food research institutions have been reduced to insignificance. If I describe some of those as "in a pathetic state of affairs" it will not be any exaggeration and in that list my old institution namely, Jadavpur University also features. The reasons of such state of affairs are many. One, – relatively inferior quality of people have come into the research and education, two, – the system has made them dull and non motivated, and most importantly, our stalwarts have not taken any pains to leave any legacy to continue.

As a result, with their disappearance from the scene the things have drifted and ultimately came to a grinding halt. And that is a major failure.

There are enthusiastic scientists like Prof. A. Sreenivasan who developed protein food by name 'Spert' – a product manufactured and marketed by Sandoz. Prof. Sreenivasan also developed the process for production of egg powder at Foods and Inns where he was a consultant.

Foods and Inns share still is quoted at Rs.60 today and the company is still doing the same thing that they used to do during Prof. A. Sreenivasan's time. When 'Spert' was launched during the late seventies I was discussing, one evening, with Prof. Sreenivasan at his Gautam Apartments flat in Bandra in Mumbai and asked him whether he really believed that there was a need of any protein food for those who can afford to buy 'Spert'. He confided in me and told me "Look Rajat if one takes full meals of only dal and rice one's proteins requirements is met and these are all humbug".

I responded by saying that "then the consumers for these so called protein foods are really taken for a ride." The marketers like us know that if there is no permanent value proposition, products do not have a long term prospect in the marketplace.

Possibly, for this reason protein foods gradually disappeared from the market like fads. Of late, I see the decline of the sales of Complain in the hands of Heinz. Prof. Sreenivasan changed jobs in public sector research institutions like one changes or sometimes forced to change in private sector.

He used to be a man of very strong conviction and a forceful character and as a result nobody liked him. But his contribution to the industry is significant.

Prof. Bose was a strong disciplinarian and perfectionist. His contribution in food research and education in noteworthy. He used to come to the department in Jadavpur University exactly at ten in the morning and used to take a round of the laboratories and classrooms.

During his time a lot of research projects were there in the department which was in the global map with respect to food research and education. Today one cannot find a teacher in the department till twelve noon.

No one comes before that. Jadavpur University did pioneering work in rice process technology, marine products processing and protein technology etc. in earlier years.

As it stands now the future development thrust for the processed food industry therefore largely has to be provided by the private sector industries itself. As the market is becoming increasingly competitive the innovations in terms of quality, cost, packaging and product variants and new products will be the name of the game.

And private sectors will have to provide the support for their own survival. This is possible. For example, Pepsi Cola in USA has kept on their roll over a dozen very senior level scientists specialised in each area of their activities to support the company's competitive strategy. This team of specialists are basically the technical backbone working for various ongoing and existing and future projects of Pepsi. In that team Dr. Shanti R. Bhowmik, a close friend of mine and a former director of CFTRI is there. Dr. Bhowmik spent 2-yr. time at CFTRI and having got fed up of the bureaucracy and the inner politics of the institution and poor level of research activities decided to return to USA again.

This was narrated to me by Dr. Bhowmik. Of course, he has also praised some of the research scientists for their quality of work and dedication and in that list the name of present Director of CFTRI Dr. V. Prakash's name featured.

I wonder if India has to be the leader in this new millennium, technical competency has to be improved for our domestic industry so that we can have more Tata-Tetly kind of organisation being created in global scenario. And that can happen only with private sector initiatives. Let us not keep much hope on our institutional and Govt.. run organisation's research effort. That has not happened in the past and I don't see that is happening in future either. The reality is that past of food research in India was more glorious than what it is now.

Acquisition of Domestic Food Business by MNCs

Liberalisation has brought in through the entry of large multinational and transnational corporations, foreign investment in this sector. This resulted in competition, technological upgradation and market expansion. In the face of the competition, domestic Industries are gradually losing market share and thus selling their businesses to the new entrant MNCs well before the value of the brand and businesses drops further due to ongoing onslaught of multinational brands.

This is happening because MNCs have much greater resources to put behind their brand and business and also have long term vision and sustainability. Domestic industries are no comparison. The first attack came in soft beverages industry with Coke acquiring Parle brands and Pepsi acquiring Dukes and with the entry of Cadbury Schweepes in this sector entire soft drink industry is now in control of MNCs.

MNCs such as Nestle, Brooke Bond and SmithKline & Beechern controlled earlier only hot beverages (both white and brown segments). But soft beverage sector was dominated by domestic industries. With Coke and Pepsi acquiring Indian businesses, total nonalcoholic beverage industry is now under control of MNCs.

Recently Sara Lee — a US multinational has acquired Nutrine — a large Indian bakery and confectionery business. Britannia is already under control of Danone group of France. Only large bakery and confectionery left to be acquired in private sector is Parle Products.

And with the entries of confectionery giants like Perfetti, Wriggly and Agrilimen (a Spanish confectioner in joint venture with Dabur but in the process of acquiring Dabur's equity to have 100% control) confectionery and bakery sector, will also become the domain of MNCs.

With Nestles' strong presence in Tomato Ketchup and Pickles and Hindustan Lever acquiring Kissan and Dipy brands from UB group through Brooke Bond route fruit and vegetable sector is also dominated by multinationals.

Heinz, the world's largest Tomato Ketchup manufacturer had acquired food business of Glaxo. Due to strategic restructuring of portfolio Glaxo decided to divest their dairy business with established brands like Complain, Glucon-D and Farex. Hindustan Lever, on the other hand, through a strategic alliance route, took control of Kwality Ice-cream as well as 100% Ice-cream brand of Jagatjit.

Hindustan Lever with their own presence in dairy sectors, and with the introduction of Baskin Robbins and Walls has emerged as a dominant player in dairy segment, in general and Ice-cream segment, in particular. Hindustan Lever (HLL) also acquired Dollop Ice-cream brand from Cadburys'.

HLL, however, recently announced their intention to sell-off their Dairy unit in UP. Ice-cream industry, which is reserved for small-scale sector, has now been in the control of multinationals through a different route. The argument that ice-cream is actually a frozen dessert — a nomenclature that has been accepted by both food and law ministry of Govt. of India — paved the way for Hindustan Lever for this strategic acquisition.

The earlier successful domestic businesses such as Dalmia Dairy, Foremost Industry and Jagatjit Industry in dairy sector are all in deep trouble — and MNCs have taken a control of this sector as well.

National Dairy Development Board (NDDB) with its strong brand Amul and with its deep root extended up to farmers level as a means of backward integration will of course, remains a significant force to reckon with. But nobody will deny that in branded dairy products Amul has to put up a constant fight against multinational brands.

Vadilal is also a significant player in ice-cream but their presence is limited in Western India and for how long they will remain has to be seen. Kraft Foods a unit of Phillips Morris is holding a licence to set up 100% owned entity and they are expected to come and set up their operation anytime.

In the convenience foods — 'Corn Products', a US multinational is fast making progress. This is an old company, which did not make any significant contribution in the Indian food business largely because of lack of interest from their principals.

CPC-International is now fast making progress with their well known Rex and Brown & Poison and Knorr soup brands. Recently they announced the acquisition of DCW Chemicals — makers of Captain Cook brands, which ran into rough weather because of the objection taken by some miller in Haryana to whom allegedly DCW. has contractual obligations.

Shortly, we will also have products in joint venture with Paoma Industries, the manufacturer of Rasna soft drink mixes and Campbell a world leader in canned foods. Godrej, on the other hand, has joined hands with Pilsburry a leading player in the convenience foods business to boost their business activities. Filsburry has already introduced instant Cake mix.

In this category others including Gits brand are relatively smaller and regional players. Apparently, first change that we have witnessed is of acquisitions of domestic brands in frontiers of food business by MNCs and alongwith although at a much lower scale a few joint ventures. Even the public sector undertaking, Modern Foods although making profit is up for sale in line with the Govt. policy of PSU disinvestment and ANZ Bank has been appointed to find a buyer for the same.

The first lesson that we have learnt as the effect of liberalisation is acquisition of successful domestic food business by MNCs.

Global Competition

Ten years ago Indian processed food industries were crying about high taxation, lack of infrastructural support, nonavailability of quality raw material at competitive price, nonavailability of latest packaging machinery and packing material etc. as the major reasons for poor or sluggish growth of this sector. But the industry was still growing then, albeit not at the expected level. Competition was also there amongst domestic processors with their MNC counterparts. Because, large MNC food companies like Nestle, SmithKline Becheern, Britannia, Glaxo, Hindustan Lever etc. are there for ages.

But domestic industries were not worried about their presence. There was healthy competition. In some areas domestic industries were doing even better than some of their nearest competitors in multinationals. What went wrong then? Why domestic industries are now not able to fight multinational giants? This is because the rules of the game have changed. Post liberalisation, we have now entered into an era of survival of the fittest.

Only couple of years ago we had only Amul Cheese. Britannia first successfully introduced Cheese from cow milk followed by the introduction of French cheese 'Le'-Bon" from Dabur-Le'-Bon joint venture.

Philips Morris — owner of Kraft Foods — is sitting waiting a formal licence to set up 100% owned company in India. They carried out lot of research for introduction of Kraft cheese in India. Their project now has been put in cold storage and they are still waiting and watching the market for processed food industry.

Upto mid eighties Kissan was the number one brand in Tomato Ketchup and around that time the competition came from Maggie — a Nestle's brand. With Maggie launching several varieties of Tomato Ketchup there was a growth in the market. With Kissan and Maggie fighting neck to neck other smaller but established brands like Dipy's from Herbertsons, Volfruit from Voltas, Noga of Nagpur Orange Federation and SunSip of Wimco gradually disappeared from the market. Delmonte now has joined hands with Paoma Industries, the manufacturers of Rasna Brands of soft drink concentrate and is expected to launch Tomato Ketchup shortly. The biggest global player in Tomato Ketchup Heinz is still waiting and watching. It was expected that they will launch Tomato Ketchup in Indian market soon. Some of the unknown local brands of course still existing only on price competition.

The price competition has also kept some of the other categories of food products still alive. For example, the age old Corn Flakes from Mohan Meakin under the brand name Champion still sells almost the same quantity (reportedly 500 MT per annum) against the much better quality Kellogs brand which sells around 300 MT per year with all its variants.

Till today we have only Parle and Britannia as two major brands in biscuits. Kellog is entering that segment gradually. Cadbury and SmithKline Becheem have nominal presence with speciality biscuits.

Shortly, we are going to see more cookies and speciality biscuits from Sara Lee, which has acquired Nutrine.

In the snack foods category we have two successful Indian players namely Uncle Chips brand of Amrit Foods and Pepi of SM Dye Chem, others were local such as Fun Flip, Crax etc. Soon entered Pepsi with their Ruffles brand and soon extended to Indian snacks by introducing bhujjias which Bikanerwala makes for them to market.

In confectionery we have over half a dozen global players struggling here for brand share and fighting against Indian players like National, Norton, Parle, Parrys etc. Amongst MNCs in this field leading are Perfetti, Wriggley, Agrilimen, Park Davis (with their chicklet brand).

Competition is also seen in organised food retail chains. Couple of years ago we had only Narula running such an organised retail food chain. Today we have everybody namely. McDonald, Kentucky's Fried Chicken, Dorninos, Pizza Hut of Pepsi's and many more. In some of these categories domestic brands are still a leader-for example, Uncle Chips, Frooty, Rasna etc. But how long they will be able to retain their leadership is a big question mark? If, potato price continues to remain at the present level possibly Pepsi's Raffle will still continue even after incurring losses but Uncle Chip will have to close its shutters.

As others are doing — for example Billy's Ltd. — an Indian company in technical assistance with Florigo of Netherlands had successfully introduced potato chips in Eastern India and was reportedly doing well. As potato price started shooting up the company could not afford to sell its brand at a competitive price and is now in the process of closing down. With highly fluctuating input price, domestic industries will be all the more vulnerable like the one as we have seen in case of Billy's.

The second lesson that we have learnt is that in each category now we have global competition with more players expected to jump into the fray. Domestic processors will have to learn to survive competition. But consumers are the real beneficiaries, they have now a wide choice.

Technological Upgradation Leading to Superior Quality

Food processing industry in India has undergone a total metamorphosis over the last six to seven years, in terms of types, variety, quality and presentation which is a result of liberalisation leading to foreign direct investment in frontiers of processed food sectors. Multinational and transnational corporations made significant contribution.

This was possible due to market significant technological upgradation in this sector. It is not that the latest state-of-the-art technologies in this sector were not known to us. On the contrary, we had the fullest capability and technical know-how.

In seventies and eighties we used to talk about in seminars and conferences the new types of packaging, products and processes but industries never felt the necessity to upgrade or invest in new technology. Nor did they have the resources to support such programme.

Reason was is the apprehension all around that there would be no market for such products. And we continued to produce products that are of traditional nature using age-old technologies. Besides, there would were licensing restrictions, reservation of many categories under small-scale sectors etc., resulting of into poor or negligible upgradation of the industry. Things have changed since then.

With the technological upgradation we now see an all round improvement in processed food products quality. The competition has forced the processors to improve and the upgrade the quality. Earlier we had limited categories, for example, jam, jellies, marmalades, squashes, cordials, ketchup and sauces in fruit and vegetable sector, ice cream, spray dried milk powder, condensed milk, butter and ghee in dairy sector, bread, biscuits and pancakes in bakery sector, hard boiled confectioneries, one brand and type of chewing gum and limited variety of chocolate in confectionery sector. Today we have not only better quality of these but also a lot of variants and product types and choices. There is also a significant improvement in packaging quality. In seventies we had only glass and cans as packaging materials. In eighties we had seen laminates and other plastic materials coming in.

Today we have the host of packaging types being used to market processed food products. Earlier we used to get fruit juices and ready-to-service (RTS) beverages only in glass bottles. Today we have tetra pack and also asceptic carton pack. Similarly, tomato paste earlier used to be sold in metal cans and the market for this was also very small. Today tomato paste is sold by Godrej Foods in asceptic consumer pack. In packaging of cheese and other dairy products considerable improvement in the product and packaging quality are evidenced. Cheese, as a product was not known to the Indian consumer in seventies and eighties. Only cheese that we had was cottage cheese.

Processed cheese was not there, not to speak of varieties of cheese as they are found and consumed in other countries. I remember during early eighties we had to develop a particular party near Delhi to supply Mozzrella cheese required for pizza to be sold through Mcdowells Fast Food chain which subsequently closed down.

Today although we don't have all varieties of cheese but we have seen a beginning. Britannia and La-bon cheese are definitely gaining ground. And we expect that when Kraft will enter this market there will be a revolution in processed cheese sector in India.

Kraft International has done considerable research on the possibility of producing world class Kraft cheese for Indian market and they came to the conclusion that the milk quality in India is not good enough to produce cheese of Kraft grade. Besides, Kraft cheese is produced from cow milk where as in India bulk of our milk is from buffalow. They even thought of importing the compressed casein block to Nepal under refrigerated condition and set up processing plant there for Indian market in order to take the custom duty advantage because quality is what they can not sacrifice.

But that also was not found to be viable. I told Mr. Puno who is in charge of South East Asian region including Indian subcontinent that starting early is the key. I advised him to start the operation first and cow milk can be made available or can be organised.

But they had still the apprehensions. Britannia subsequently introduced cheese and their sole claim is "Cheese from cow milk" and has done very well if initial responses are any indication. And it is everybody's knowledge that Britannia is not manufacturing it themselves but it is being manufactured in a local plant near Pune under a subcontract manufacturing arrangement.

Britannia is only marketing it. Although process technology support might come from their respective French collaborators namely BSN and LA-BON, but this only proves my contention that technology is known to us but we did not put it to application to tap new and emerging market. Competition is forcing the marketeers to look for better products and new products, which is a direct outcome of the liberalisation policy.

Even the traditional item like Indian snacks (bhujjia) and Bengali sweets manufacturers are also thinking to upgrade their manufacturing technology and product quality. Mr. Agarwal, Managing Director of Bikanerwala who supplies bhujjia to Pepsi Foods recently invited me to visit his plant at Lawrence Road in Delhi and said Pepsi's global chief also visited his plant and was highly impressed to see their process.

When asked what exactly the improvement they have made in their age-old labour oriented process, Mr. Agarwal commented that the packaging of Bujjia is now fully mechanised.

Similarly sweets are also a serious trade worth Rs.16500 crore and in West Bengal alone the market is estimated to be Rs.1000 crores. The Sweet Traders Association in West Bengal recently held a seminar in Calcutta to devise ways of making the sweet making process more hygienic and to ensure optimum use of technology _ said Mr. Robin Pal, General Secretary of the Association. The implications of liberalisation has also reached these so-called traditional foods and the traders now are eyeing for the global market.

Under liberalisation regime consumer is the king. They are getting better value for money and they have now a wide choice to select from.

Proliferation and Extension of Food Brands

Food items are primarily processed to extend its shelf life and to preserve its nutritional value. The secondary purpose of processing of food is to deliver convenience, taste and appeal to the target consumers. Depending upon what consumer segments we are targeting the processing methods, type of food and its taste and packaging and total presentation varies widely. In a country like India where local taste preferences differ from region to region it poses a much bigger problem to produce food products which can have a much larger acceptance and hence a bigger market potential.

But there are some food categories and types for which there is a universal acceptance and hitherto Indian processed food industry only were processing those generic category of foods for all India market. Even new entrants in the processed food industry were only adding brands in the traditional sectors. We have witnessed this phenomenon before liberalisation in different categories. And as a result, mortality rate of processed food brands were quite high. For example, when Amul came into the market with their Amul Butter-Polsons butter disappeared from the scene.

And the successful brand went on extending it to other categories. Having tasted success, Amul brand has been extended across a range of items including cheese, cheese spread, shrikhand, chocolate and ice-cream. Creativity in food industry was, to say the least, very low. There were of course small local companies who attempted to create products for catering to the local taste preferences but the market was so small that the products did not survive the test of time. For example Kissan launched "Bisebelle Bhath" a South Indian delicacy in Southern market in mid eighties only to withdraw a year later.

We at Kissans did considerable work and developed instant rice but market research indicated that the market for instant rice would be insignificant and thus we had to drop the idea.

The same instant rice was used to formulate "Bisebelle bhath". Phillips Morris who got clearance recently for setting up 100% owned entities in India has announced their plans to introduce the instant rice under their global and popular brand "Minute Rice".

And still there are others who tried to create universal appeal and introduced mass market product and have demonstrated big success stories. The example is Rasna from Pioma Industries. But how they did it is a different story and that will be covered in another article.

Yet others have failed in such attempt. In mid eighties steel merchant of Mumbai Mr. Ravi Kiran Agarwal in a joint venture with Kwality Frozen Foods had set up Tasty Bite Eatables Ltd.-a company to get into frozen foods and North Indian and South Indian vegetarian and non-vegetarian ready-to-serve foods.

The concept was-warm and serve. A modern plant was set up near Pune. The products including Mutton curry etc. in retortable pouch were introduced but it did not click. A couple of years later they faced serious financial problems and subsequently the company was sold to Hindustan Lever-who tried hard to establish "Green Valley" frozen foods. It was also later on withdrawn.

Welga Foods in Badaun near Bareilly similarly had to close down their frozen foods plant. Tarai Foods who were also contract manufacturers for Hindustan Lever is still struggling to exist. The only long term player in frozen foods particularly frozen vegetables is "Mafco" — a co-operative in Maharashtra. Food products that were in the high price range and were well ahead of their time did not succeed in India. And thus casualties in certain category were very high.

Before nineties the food industry in India was struggling to establish only the traditional food brands and there was very little or no effort to introduce new foods.

In post liberalisation era we have host of new MNCs with superior marketing skill and bigger resource to their command have extended their brands to a wide variety of local foods and their variants including the areas which were protected niches of small local players. For example, Indian traditional snacks were only local brands even a couple of years ago. But there we have now even global brands. Kissan under UB-group used to sell and market only fruit and vegetable products. Today Kissan brand has been extended to even branded grocery items which were earlier the domain of small local players.

Pepsi's primary business is in soft drinks but in India they have gone into branding of traditional snack foods. Similarly Nestle-a swiss

Food major-has introduced pickles in Indian market. In the traditional food items local companies should be able to add better value and hence can score higher in the market place.

In seventies, 'Bisleri' was the only mineral water which had national presence and the sale was to the tune of approximately one hundred thousand cases valued at about Rs.60 lacs. Today mineral water market is said to be of Rs.200 crores and over hundred brands of mineral water are there.

Mr. Ramesh Chauhan of Parles thinks that mineral water market in next five years time will become Rs.2000 crores and based on that he has drawn up ambitious plan for his Bisleri mineral water.

The competition in mineral water is so high that Mr. Ramesh Chauhan's own brother Mr. Prakash Chauhan who owns "Frooty" brand also introduced a mineral water under the brand name "Bailey". Of late we have multinationals entering the mineral water scene. "Evan" is recently launched which is a well known international brand.

Earlier we did not have any fruit juice in the market. Only canned juice and ready-to-serve beverages in bottles were available. Today we have number of brands in carton and tetrapack. Such as 'Onjus' of Enkay Texofood, 'Real juice' of Dabur, 'Jumpy' of Godrej etc. all packed in convenient asceptic cartons. We will be shortly getting more brands of fruit juice including 'Tropicana' from Seagram.

The proliferation of brands are also seen in branded Grocery products. A few years ago we had 'Tata-salt' as the only refined iodised salt selling in the market and today we have 'Captain Cook' from DCW, 'Kissan Annapurna' from Kissan Products and Catch (with dispenser) from Hi-Tech Foods and more brands to come in.

In some areas such as meat and poultry products branding, packaging, processing and distribution on a national scale has yet to be witnessed. But in many of the processed food categories, as we can see that the existing brand is either extended to a wide variety of foods such as Britannia to Cheese and Long Life Milk in tetrapack etc. or there are proliferation of brands.

In this process, ultimate beneficiary is the consumer as they can expect to get the best value for their money. Competitive pressure will make that possible.

MNCs' Entering Traditional Foods Sectors

Mckensey did a study on the opportunity in India for food and agri business at the behest of Confederation of Indian Industries (CII).

And as expected they projected a huge potential in food sectors and said that the real potential lies in traditional, local food and grocery items — but not so much in westernised processed food. Some of our big multinationals seem to have swallowed the Mckensey advise-pill and got into this sunrise sectors.

Historically, traditional food items were the domain of small local players. Grocery items earlier were sold loose and unbranded. Then some progressive traders started the cleaning, grading and primary packaging. And the next stage of development was branding of primary foods such as Atta (wheat flour), Rice etc. but those were mainly regional brands. Popular among these were "Rose" and "Shakti Bhog" brands in northern India.

Whereas in traditional food items such as spice and pickles there were host of local brands with regional strong-holds such as 'Cookme' spice powder in East, 'Bedekar' pickles in West, 'MDH-masala' in North etc., to name a few. And the food ingredients — nobody thought of. Only food ingredient that was there was tomato puree and paste in cans which in the recent times Godrej introduced in tetrapack.

Edible oil in tin pack and later on in polyester jar and tetrapack were the first major step in branding grocery items. "Postman"— peanut oil from Ahmed Mills was one of the oldest popular brands.

Today we get variety of edible oils and NDDB and Marico are the leading players in this sector. Amongst MNCs we have "Mazola" corn oil from CPC and "Sundrop" from ITC Agro — a sunflower oil which subsequently has been acquired by CONAGRA-the fastest growing food company in America. With CPC and CONAGRA being involved in edible oil market this sector is going to be soon dominated by MNCs.

The next major attempt to market a branded grocery item was done about a decade ago by Tatas by introducing the first refined iodised salt in poly bag by brand name "Tata Salt" and that was a success. The next big brand in this category is "Captain Cook" from DCW Chemicals who had plans in branded grocery items and thus the company introduced other items including wheat flour under the same brand name.

But that was shortlived. During 1997 DCW Chemicals suffered a major setback and the company decided to divest this food business. The acquirer is Corn Products Co. (India) Ltd. — a wholly owned subsidiary of CPC-International a nine billion US dollar American multinational in food and grocery business.

The company has recently changed its name to Best Foods Ltd. CPC was a sleepy company operating in India for over fifty years and their performance was lacklustre. With their limited range of Rex and Brown and Polson brands of convenience foods, the parent company, although doing well globally did not have any commitment in India.

In late seventies and early eighties when I was heading their project department many investment proposals were sent and notably among those were dextrose manufacturing project as CPC has a flavoured glucose brand in the portfolio and Knorr soup project. But project work used to be carried out by us endlessly but nothing got implemented.

In post liberalisation era, this apparently sleepy company seems to have become active. The third brand of salt is "Kissan Annapurna" which is making a sustained effort to get market share. This brand which is now in Hindustan Lever's fold through Brook Bond acquistion who in turn acquired Kissan in early nineties have reportedly spent Rs.18 crores in advertising and got 14% market share in a Rs.200 crore branded salt category in a span of one year.

Tata Salt still holds 26 percent market share and Captain Cook's share is about 20 percent and balance forty percent is still with the smaller players. With the acquisition of "Captain Cook" in branded salt market MNCs are the dominant players controlling sixty percent of the market.

The value addition in branded salt and pepper was really done successfully by a Delhi based local company Hi-Tech Foods belonging to Dharampal Satyapal of premium "Baba Zarda" (chewing tobacco) fame. Hi-Tech Food's `Catch' brand of salt and pepper in dispenser pack is a success story with upper middle class household and restaurant segments as the main customers. The sale of "Catch" is said to be around Rs.12 crores.

The first MNC to get into the branded spice business was Brook Bond who introduced the select premium priced spice range by name "Sona" in late eighties. The products were priced and packaged for higher income group.

They struggled for couple of years to establish "Sona" brand but failed and was forced to withdraw in later years. Apparently we don't see any major MNC in branded spice but world leader in exotic spices "Macomac" seems to have plan to enter Indian market. The first MNC to introduce Indian pickles was Nestle with their Maggie brand.

The products are still in the market. And as mentioned earlier, the first MNC to introduce traditional snacks (bujjias) is Pepsi under the umbrella of "Leher Namkins". It can be concluded that in traditional Indian foods MNCs cannot add much value through the involvement of their principals abroad. On the contrary, they will have to learn from the locals to derive advantage of their brand and resource muscle.

Hindustan Lever is expected to enter branded grocery items though Kissan route in a big way. The company does significant export of branded rice particularly in Middle East market and therefore expected to introduce the branded rice shortly in domestic market as well. Otherwise, in India so far branded packaged rice is limited to Basmati rice and there are many brands in this category and leading among them is "Kohinoor" — but all are from small manufacturers.

The grocery business is basically low margins and high volume business and thus it required multiple supply source to be strategically located to reduce the cost of freight and excellent distribution infrastructure and logistics management capability. Big houses like HLL, therefore are expected to do much better in this sector.

In the food ingredients business MNCs have not yet traded. The only big player in this area is Dabur with their range of pastes such as onion, garlic, ginger and also mix of these for particular application.

Current sales of these items which are sold under the umbrella brand "Hommade" is said to be Rs.5 crores and going by that standard it can be considered as a reasonable success. Very recently, "Cookme" curry paste has been launched in Calcutta. The market indicators point towards growth in the branded food ingredients business in coming years.

Pepsi is expected to get into this category and with their backward linkages with farmers whom they support with the high yielding variety of seeds for better crop yield, it stands to have an edge over others in this sector. One could have expected that Nestle would also jump into the fray in this sunrise food business — but their new Italian Managing Director Mr. Carlo Donati seems to have other priorities and they want to focus on major key brands.

Price and quality are expected to be the major determinant in the success of grocery items and basic foods. Consumers in India would not be willing to pay much as the price of convenience. Big global brands to MNCs' in that context are therefore not much of significance.

Imported Foods in Indian Market

Not too long ago for buying imported foods we had to go to select markets-such as Modern Bazar. Vasant Vihar in Delhi or Fort in Mumbai. In other places one has to literally search for such choice food items. In those days importing foods were not permitted.

And we were told that these foreign made chocolates and soup mixes and cheese find their way to Modern Bazar through diplomatic routes in the sense that persons in diplomatic services bring with them these foodstuff and sell them to local stores. But the fact of the matter is that these food items are brought to India through what is called "Suitcase trade" in the concerned circle. It will not take time for readers to understand what the "suitcase trade" is? It simply means to carry suitcase full of foodstuff by people frequently travelling overseas and selling and or delivering those to local traders and this happens in an organised manner. Those who can afford, have to pay a fancy price for those foreign foods. Although price was exhorbitant but the quality used to be distinctly better. Children love their Swiss chocolates, French Cheese and Italian pastas.

General Foods (now known as Kraft Foods International)-subsidiary of Phillips Morris has a synthetic orange beverage by brand-name "Tang" found to their dismay that so much of Tang is selling in India and Nepal now that they started thinking what would be the loss of sales to their South East Asian subsidiaries from where it finds its way to Indian market, once they start manufacturing it locally.

In the last couple of years these imported foods are available almost in all leading shops in most of the shopping centres in the city. To my surprise while taking a stroll in Greater Kailash shopping centre, I found couple of those shops have stacked Champaign, sparkling wine and imported mineral water outside the shop alongside soft drinks and on getting curious when I peeped inside I found everything is available. You name it — it is there. I bought some MARS chocolate and came out. The way things are moving, some of these foreign companies in course of time will do only trading in foods.

There are local agents who now represent many foreign companies and regularly import foods in turn to supply to local retailers and stores. In big stores like Nanz, there are regular requirement being catered to by their local agents.

In earlier years we used to see these phenomena in countries like say in Middle East. This trade in imported foods will be on the increase as the custom duty goes down. It will also fuel the growth of partial processing in India to take the advantage of price equation. For example, importing the product in bulk and carrying out re-packing operation here will be more economical than importing the finished product itself.

As we know that India has now signed the WTO (World Trade Organization) agreement, of course, being a signatory to this agreement is a lesser evil than not signing it at all. As per the terms of reference to this agreement, by year 2003, India has to put most of the items under Open General License (OGL) for import which means these items can be freely traded. We so far could announce around about hundred odd items under OGL and in that list shrimp, strawberry, rasberry and other fruits, have already featured. With the recent announcement of the Commerce Ministry, we have now freed import of 894 items including consumer goods and agricultural products by putting them under Open General Licence (OGL).

Another 414 items are removed from the "restricted" list allowing these to be imported against Special Import Licence (SIL) that can be procured in the market by importers at a small premium. Fruits like oranges and pomegranates, soft drink concentrates etc. are falling in SIL category whereas a range of marine food products, vegetables, assorted fruits and juices, spices, edible oils, jams, jellies etc. are brought under OGL.

The peak rate of the basic import tariff is now 40 percent which is applicable to most consumer products. On some items such as milk products India's WTO commitment bind it to a zero duty. But in most other agricultural products the WTO restriction is substantially higher ranging from 150 percent in the case of processed foods to 300 percent for edible oils. It is interesting to note that India, as a part of its globalisation policy is reducing the general tariff structure much faster than the asking rate as per WTO.

The outcome of the same is that in India we now get "KIWI" — an Australian fruit from select outlets if we want to. Shrimp is also now imported into India.

In a recent wealthy marwari marriage reception that I had attended, I found in the fruit counter all the fruits look different than what I am familiar about. For example, Guava and Apples were almost double the normal size that we get here. Plus there were unknown fruits. And I was told that all these fruits were imported from Bangkok.

I was so tempted that I tried all varieties and filled my stomach with all these tasty exotic fruits only. And later on started thinking that what will happen to our fruits if these fruits are available at a competitive price in India. Can our farmers and fruit merchants become globally competitive in terms of price and quality.

We should see the writing on the wall and work in that direction. The Agricultural Products Export Development Authority (APEDA) should quickly assess the scenario and rise to the occasion to take appropriate action. All of us are seeing the ongoing battle European banana growers are fighting in the forum of WTO against Americans refusal to permit entry of European Banana in that country.

If international fruit brands like "Chikita" and "Del Monte" start selling here, who is going to buy local fruits from our mandis and roadside vendors selling fruits on the top of hand-pulled trolleys?

Food Forum India: Redefining the Indian Food Business

The Food Forum India, the largest congregation of key, global and Indian, food retailers, manufacturers, organisations and minds in the food business will be held from May 6-7, 2008 at Renaissance, Mumbai. It will span the vast array of food business segments like food retail, food design, food processing, food logistics & support, food services, information & technology.

Conceptualised at the India Retail Forum (IRF), which saw the participation of leading brands, retailers and retail support organisations from across the globe, Food Forum India (FFI) was launched under the chairmanship of Subodh Kant Sahai, Union minister for food processing industries, with some 50 captains of the food world joining in to take up a series of support initiatives for the Indian food businesses.

The exhibition space covers all aspects of the food retail supply chain and has different exhibition zones such as Food Bites, Beverages, Food Support, Food Design, Food Space and Food Concepts.

F&G Retail Projections For 2010-11

- Reliance Fresh: To cover 70 cities, 784 urban towns and 6,000-odd rural mandis
- Future Group: 500 Spencer's Fresh, Daily, Super and Hyper
- Wadhawan Group: 1,500 Spinach/S-Mart/Sabka Bazaar outlets in 100 cities
- Subhiksha : 2,500 neighbourhood outlets across the nation

- Other major players firming up Pan-India plans include Bharti-Walmart, AV Birla Group (Trinethra/More), Heritage Foods (Fresh @, Hypercity, Dairy Farm, (Foodworld), Nilgiris, Radhakrishna Foodland, Influx, Landmark Group (Spar), Mother Dairy and Safal & region specific expansion by Jubilant Group (Jumbo Saver & Monday 2 Sunday), Hopcoms and Margin Free in South, Adanis and Arambagh Foodmart in East India among others. Most major retailers are expected to foray in to cash and carry model, to structure backend and take full advantage of large scale production, procurement, processing and also to put supply chain & logistics mechanisms in place.

Conferences : Several conferences will be held during the event.

Food Inspirations : The ability to do things differently has been one of the prerequisites to large scale success. Captains of successful food businesses will share their experiences and give insightful interpretations of the food industry.

Food Innovations : New ideas, products and concepts. Discussions on some of the most important innovations that have gained maximum popularity over the years. A look at possibilities for the near future.

Food For Thought : Presentations of leading global research and studies that are shaping the food industry today. Leading global food research and consulting firms and organisations will deliberate on key issues that guide the food business.

World Food Experience : The global market is today witnessing a mushrooming of numerous ideas and innovations that are radically altering the shape of businesses around the world. Focus will be on international ideas and insights, and include sunrise products from around the world and their implications for the Indian market.

Food Fight : Interaction between retailers and producers is of great importance to ensure the maintenance of an efficient chain that facilitates development of the industry as a whole. Food retailers and producers will debate on vital issues to determine the way ahead for the food industry in this session.

Investments In The Food Sector : Trade commissioners of key countries, governmental and regulatory bodies, international and domestic investors and corporates will debate upon investments in food processing, retailing, support and infrastructure.

Future Of Food In India 2020 : National and international vision leaders and key players of the industry will foresee the future of the business of food in India, and present some of the astounding possibilities of the future.

Awards : The Images Golden Spoon Awards (IGSA) to be announced at the event will honour enterprise, innovation and achievement in the business of food. Industry captains will celebrate excellence in the food business at a glamorous ceremony that will be the grand finale of Food Forum India.

The IGSA awards will be benchmark of excellence, that all food businesses will aspire for, thus playing a vital role in the in the progress of the industry. Food and grocery accounts for 63 per cent share in the Rs. 1,330,000 crore (US$ 300 million) Indian retail market. Only one per cent of this segment was organised in 2007. Food, grocery and beverages constitute about 43 per cent of India's private final consumption expenditure, and yet only 11 per cent of organised retail market in 2006.

Opportunity in Food and Grocery

The Urban Middle Class continues to grow by 20-25 million annually. The population is young, better exposed to modern lifestyles, and has a preference for packaged food and beverages. This growing Indian Middle Class, especially the employed youth in the ITES, BPO and other Service sectors, have higher propensity to consume, fuelled by easy availability of credit due to popularization of plastic cards.

More than 1,000 Food and Grocery outlets operational across tier-I and tier-II cities and Retail space of over 3 million sq. ft. have come up for F&G supermarkets, department stores and hypermarkets. 25 major F&G Retailers emerged in less than a year and Retail sales in this segment grew by nearly 150 per cent over 2006.

Opportunity Logistics

Improved logistics, warehousing, state-of-the-art technology and availability of quality processed and packaged food are factors that will boost the growth of organised F&G retail in India. Indian corporates like ITC, Godrej, DCM Shriram have initiated the process of direct supply linkages with the farm. Global major Metro AG was the first to introduce up-to-date know-how logistics and supply chain management. Corporate major Reliance is investing heavily on this front. Wal-Mart too plans back-end operations with Bharti.

Investments to the tune of US$ 23.5 billion in the food processing sector over the next three years is in the pipeline. This will undoubtedly create a huge demand for the food processing and packaging industry.. But keeping in view the vast farm sector and the surmounting demand for quality food products, there remains room for much more.

Opportunity in Food Processing

Nearly 80 per cent of agricultural produce in developed countries gets processed and packaged. It is just about 1.3 per cent in the case of India-a reason why less than one per cent of the F&G retail segment is organised. Even the existing organised F&G Retailers are finding it difficult to keep their stacks full & fresh.

This is just the right time for global firms to invest in the food processing sector, especially so when the government seems eager to boost rural/farm incomes. Marine Food Parks and Agricultural Special Economic Zones are already being worked out to ensure that farmers are not displaced. The opportunity is huge.. and early movers will reap the maximum advantage.

Opportunity to Boost the Incomes of Indian Farmers

Organised F&G retail requires direct sourcing of bulk produce and this ensures a boon for farmers. Initiatives thus far, are proving the point as traditional mandis loose relevance, middlemen get eased out, and as a result there is substantial increase in the income of farmers.

Adverse effects of fragmentation of land holdings can also be effectively countered. Initiative and Investment by Global companies can help make Cluster farming, Demand oriented farming and Raw material oriented farming the mainstay for Indian farmers.

India could well become the agri-business powerhouse in the export of fruits, vegetables, spices and food grains.

Opportunity for Food and Farm Supplies

More than 5,000 new outlets, 100 hypermarkets, 500 department stores and 2000 supermarkets are opening. Over 5,000 small and big existing outlets are to undergo a complete facelift. But the supplies are inadequate. Organised retail in India will need fresh and processed food and farm produce worth US$ 30 billion by the year 2010.

F&B (Servicing) Retail in India

Home-grown as well as international restaurant chains present in both high street locations and malls represent the organised food and beverages (F&B) services retail sector. This sector is in its growth phase and offers opportunities across a variety of retail formats.

The Indian F&B services sector is estimated to be worth Rs. 70,000 crore in 2007, at current prices, out of which about Rs. 5,750 crore is accounted for by the organised sector.

This is roughly 8.2 percent of the market being organised as compared to just 6.9 per cent of the market being organised in the previous year. This F&B services sector has increased its share from 4.7 per cent of the total of Rs. 1,200,000 crore retail market in 2006 to about 5.26 per cent of the Rs. 1,330,000 crore retail market in 2007, as per preliminary estimates by Images F&R research.

Report – India's Food Processing Industry

The Food Processing Industry sector in India is one of the largest in terms of production, consumption, export and growth prospects. The government has accorded it a high priority, with a number of fiscal reliefs and incentives, to encourage commercialisation and value addition to agricultural produce; for minimising pre/post harvest wastage, generating employment and export growth.

Important sub sectors in food processing industries are:-Fruit & Vegetable Processing, Fish-processing, Milk Processing, Meat & Poultry Processing, Packaged/Convenience Foods, Alcoholic beverages & Soft drinks and Grain Processing etc. As a result of several Policy Initiatives undertaken since liberalisation in August 1991, the industry has witnessed fast growth in most of the segments. As per a recent study on the food processing sector, the turnover of the total food market is approximately Rs.250,000 crores (US $ 69.4 billion) out of which value-added food products comprise Rs.80,000 crores (US $ 22.2 billion).

Since liberalisation in Aug 91 and up-till Feb 2000 proposals for projects of over Rs.53,800 crores (US.13.4 billion) have been proposed in various segments of the food and agro-processing industry. Besides this, Govt. has also approved proposals for joint ventures, foreign collaboration, industrial licenses and 100%export oriented units envisaging an investment of Rs.19,100 crores (US $ 4.80 billion) during the same period. Out of this, foreign investment is over Rs. 9100 crores (US $ 18.2 billion). Processed food exports were at over Rs.13,500 crores (US $ 3.2 billion) in 1998-99. Out of these exports, rice accounted for 46%, whereas marine products accounted for over 34%.

Primary food processing is a major industry with lakhs of rice-mills/hullers, flour mills, pulse mills and oil-seed mills. There are several thousands of bakeries, traditional food units and fruit/veg./ spice processing units in unorganised sector. In the organised sector, there are over 820 flour mills, 418 fish processing units, 5198 fruit/ veg processing units, 171 meat processing units.

India is the world's second largest producer of fruits & vegetables, but hardly 2% of the produce is processed. India is the land of spices producing all varieties worth over Rs. 3500 crores (US $ 900million) amounting to 25-30% of world production, which is processed for value-addition and export. It grows 22 million tonnes of oilseeds covering most of the varieties. Other important plantation products include tea, coffee, cocoa and cashew.

It has large marine product and processing potential with varied fish resources along the 8041 km. long coastline, 28000 km. of rivers and millions of hectares of reservoirs & brackish water. India's livestock population is largest in the world with 50% of world's buffaloes and 20% of cattles, but only about 1% of total meat production is converted to value added products. India is the largest milk producer in the world and about 15% of the total milk production is processed through the organised sector. Size of the semi-processed and ready to eat packaged food industry is over Rs. 4000 crores (US $ 1 billion) and is growing at over 20%.

With India's food industry witnessing explosive growth in line with the country's on-going economic development, there has been a marked move towards trying to get more out of the country's immense agricultural industry. Despite being the world's leading producer of cattle, milk, tea and pulses and the second largest producer of fruit, vegetables, rice, wheat and sugarcane worldwide, India accounts for just 1.5% of global processed food trade. The government estimates that the cost of boosting this figure to just 3% would be US$25bn, with India's primary focused agricultural and food supply chain having to dramatically reconsider standard industry practices in order to boost secondary output.

Fortunately, the government is not alone in attempting to fund this seismic shift. A vast population, steady urbanisation and GDP growth that is forecast at 7.9% annually to 2012, have helped ensure that multinational investment interest in the market is massive.

More encouraging than this is that multinationals alone are not being left to drive development; local corporations are also playing a huge role in the process. Mos notable has been the involvement of Indian companies in the food and beverage industry, whose primary business interest was formerly elsewhere. Over the past 12 months this has seen fast-moving consumer goods manufacturer Hindustan Lever – 52%-owned by Dutch giant Unilever – attempt to grow its food division, via the heavy marketing of popular brands, such as Knorr and Lipton, and reduce its dependence on its health and personal care products.

Perhaps even more significantly than this, burgeoning food sales have seen cigarette industry number one and two – ITC and Godfrey – launch new food products in a bid to establish themselves as key players within this industry.

Looking at the industry more broadly, almost all of India's leading conglomerates – Tata, Reliance, RPG, Aditya Birla etc. – are now participating in the industry to some degree, many having chosen the explosive grocery retail as a starting point. Likewise, former food category specialists, such as Amul Dairy, are now looking to become industry generalists in a bid to broaden their growth prospects (Amul has recently announced its interest in further non-dairy snack food launches).

Investment from all of these companies, as they battle to boost sales and build market share, should help to lift India's contribution to global processed food trade in the long term. In the short-to-medium term, the impact will be on internal processed food sales. Included in this report for the first time this quarter are our forecasts for the country's processed food industry. To 2012, we expect processed food output to increase by 44.2% to US$90.1bn, while packaged food sales in the country climb by 67.5% to US$21.7bn. On a per capita basis, it is forecast that a per capita packaged food spending increase of 56.5% to US$18.06 in 2012.

Food Processing – A New Era in India's Economy

The agriculture sector has come a long way since independence. With the advent of green revolution, India has transformed itself from a country of shortages to a land of surpluses. With the rapid growth of the economy, a shift is also being seen in the consumption pattern, from cereals to more varied and nutritious diet of fruit and vegetables, milk, fish, meat and poultry products.

This has resulted in the development of a sunrise industry namely the Food Processing Industries. The food processing sector in the country with its vast potential has emerged as one of the major driver of economic growth. It is encouraging to note that while the country's GDP growth rate had increase from 3.5 per cent in 2002-03 to 9 percent in 2006-07, the food processing sector has grown from 7 per cent to 13.1 per cent during the same period.

India is a country of over 1.10 billion consumers, there is a large untapped domestic market of 1,000 million consumers in the food processing sector and 200 million more consumers are expected to shift to processed food by 2010. It is the second largest producer of fruits and vegetables in the world.

There is a huge wastage of perishable food items in the country due to lack of proper food processing facilities and the level of processing is only about 2.2 per cent. However, India has tremendous potential to unleash large scale process based farm activities to exploit the emerging globalbusiness opportunities.

Incredible Opportunities

India's homogeneous market size endowed with growing incomes and changing life styles has created incredible market opportunities for food producers, machinery makers, food technology and service providers. Food processing industries has great export and employment potential. The policies are investor-friendly and more importantly technological and human resources are available aplenty in the country.

The competitive edge enjoyed in terms of raw material and labour offers lucrative opportunities. However poor perception of quality and the indifferent image of Indian products is preventing Indian food products to penetrate global markets in a big way. While developing countries like Thailand have exploited the global markets in a big was by fine-tuning quality management aspects of their food processing industry, India are yet to make a headway on this front.

Production of high quality processed foods meeting international quality standards & regulations may very well open new frontiers for Indian food products. This will not only create a dynamic and competitive domestic food processing industry but will also enable India to become a major player in the global food market. An attitudinal change towards quality is essential. Several thousand crore worth of farm produce is lost every year due to inefficient postharvest practices for storage and processing. On one hand is the growing demand for food products, which are difficult to meet due to limited resources and on the other, there exist abnormally high wastage in farm sector due to inefficient technology in storage, processing and handling.

It is, therefore, imperative to introduce state of the art technology in the food-processing sector to minimize post-harvest losses. It also calls for a concerted attention to a few selected food products where India has or can develop a competitive edge over other countries.

FDI

The sector has been attracting substantial FDI also and is among the top ten sectors getting FDI equity. FDI up to 100 per cent equity is permitted under the automatic route in food and infrastructure like food parks and cold chains. There are many areas for investment in this sector which include mega food parks, agri-infrastructure, supply chain aggregation, logistics and cold chain infrastructure, fruit and vegetable products, animal products, meat and dairy, fisheries and seafood cereals, consumer foods/ready to eat foods, wine and beer, machinery/packaging.

Productivity and Progress

It is essential to understand the dynamic relationship that exists between productivity and progress. The basic fact is that until both the farmers as well as the processors are convinced of benefits that accrue through productivity, the productivity campaigns will remain ineffective. So the main challenge is to introduce the concepts of productivity and make it work under a variety of constraints for the sustainable growth of the industry. In the process of globalization, the Indian food processing industry will be facing increased competition, particularly in domestic markets in addition to the uncertainties prevailing in the international markets.

It is in this context that emphasis must be given to improve productivity and quality. Undoubtedly, better performing firms will have a competitive edge over others. In order to maintain the tempo of productivity and quality, the National Productivity Awards have thus assumed much greater significance. Ministry of Food Processing Industries on its part, is leaving no stone unturned to achieve the multiple objectives of stepping up the growth, higher farmer income, reduction in wastage, providing nutritious and safe food and enhancing employment opportunities.

It has initiated measures to deal with the major constraints being faced by the industry such as affordability and cost of processed foods, linking of farmers and processors, supply chain and post harvest technology, infrastructure, finance, food safety, hygiene and taxes. With the active support and cooperation of all the stakeholders the ministry is confident of providing the necessary momentum for the rapid growth of the food processing sector and usher in a new era in the Indian economy.

Vision 2015

Given the strengths and opportunities of food processing sector, a Vision 2015 has been developed by the Ministry of Food Processing Industries, together with an appropriate strategy and implementable action plan so as to enhance farmer income, generate employment opportunities, provide choice to consumers at affordable price and contribute to overall national growth by increasing: the level of processing of perishables from 6% to 20%, value addition from 20% to 35% and share in global food trade from 1.5% to 3%.

The vision 2015 of the Government for the food processing sector aims at enhancing and stabilizing the income level of the farmers by assuring wider and better choice by enhancing dynamism, competitiveness, by ensuring safety and quality of food by introducing a transparent and scientific system of standards.

To achieve these aims a transparent and industry friendly regulatory regime is proposed to be established. Making the sector attractive for both domestic and foreign investors. Achieving integration of the food processing infrastructure from farm to market.

Having a transparent and industry friendly regulatory regime. Putting in place a transparent system of standards based on science. To achieve the stated Vision, Ministry of Food Processing Industries, has prepared Action Plan for energizing the food processing sector during 11th Plan.

Main Initiatives

Cold Chain: To address the situation and with a view to create a modern cold chain for preservation and value addition of perishables, during the 11th Plan, the Ministry is launching a revamped comprehensive Cold Chain Infrastructure Scheme for creating integrated cold chain infrastructure at different levels – farm level primary processing centre-cum-cold chain, collection/aggregation centres and Strategic Distribution Centres (SDC).

The SDCs will have integrated infrastructure facilities like material handling equipment, refrigeration, IQF/Blast freezing facility, Frozen/CA/MA Storage, Modern Packaging Facilities, ancillary equipment like X-ray, weigh bridge etc. The SDCs will be linked to retail supermarkets.

Mega Food Park : A new scheme of Mega Food Parks in the country is proposed which is envisaged to be a well defined agri/ horticultural processing zone containing state of the art processing facilities with support infrastructure and well established supply chain. The proposed scheme aims to provide a mechanism to bring together farmers, processors and retailers and link agricultural production to the market so as to ensure maximization of value addition, minimize wastages and improve farmers' income. The Mega Food Park is designed ultimately to link the farmers with the retail markets with minimizing of the intermediaries.

These food parks will function as sourcing hubs for the retail outlets. Abattoirs: Hygienic and scientific slaughtering as well as optimum utilization of byproducts are issues of grave concern of the Indian Meat Industry. It results in tremendous waste, contamination and avoidable cruelty to animals. Ministry is launching a comprehensive scheme for modernization of existing abattoirs/establishment of modern abattoirs at 100 locations across the country on a PPP mode.

Capacity Building: Ministry of Food Processing Industries has also taken up quality assurance, R & D, HACCP, Human Resource Development and Establishment of laboratories to support the Food Safety and Standards Act. Ministry of Food Processing Industries have taken many steps to give impetus to this sector which include virtual delicensing of the sector, inclusion in the priority sector for lending, allowing 100% FDI except in alcoholic beverages and retail, several duty and tax reliefs, financial assistance for infrastructure building, setting up of food processing units etc.

In case of export-oriented units, foreign investment is permitted even in case of items reserved for small scale sector. In addition, the export oriented units are given a number of incentives and concessions under the Export-Import Policy, such as, duty free import of capital goods, raw materials and intermediates, export income being exempt from Corporate Tax etc. FDI inflow in food processing is becoming stronger.

Recipe for Success

The global packaged food and nonalcoholic beverages industry, with more than 30 distinct product categories, is complex and highly fragmented—and worth about $2 trillion,or the rough equivalent of the GDP of the United Kingdom. The top 20 players account for less than one-quarter of the market, which is characterized by vast differences in local taste preferences, shelf-life considerations and food technologies.

All of these factors have combined to inhibit consolidation. However, deal activity has picked up in recent years as more companies have sought focus by trying to unravel diverse and undifferentiated portfolios. Yet even with increased consolidation, most companies are struggling to sustain margins.

Rates of growth and profitability vary significantly across the 30 product categories. Sales growth of bottled water far outstrips that of carbonates, for example. And sales of ready-made meals are almost double those of pasta. Such differences reflect rapidly evolving patterns of consumer demand—a reflection, in turn, of profound changes in demographics and lifestyle. More women in the workforce and increased urbanization have fuelled worldwide demand for convenience foods. At the same time, the rates of obesity and food-related diseases like diabetes have skyrocketed, powering a parallel desire for "healthy" food as well as a similarly health-related boom in organic foods, especially in developed countries.

Meanwhile, as consumers in many countries become wealthier, more informed and thus more confident, they have developed an interest in eating for pleasure and have cultivated a taste for exotic, ethnic foods. There are, to be sure, significant differences globally in these trends. Interest in healthy nutrition rises in line with disposable income, while the growth in convenience consumption is a function of both rising incomes and the increased sophistication of local retail networks. Emerging markets offer sizable growth opportunities, if only because of their relatively low per capita consumption rates.

The bulk of future growth, however, will come from North America, Western Europe and the Asia Pacific region, where the impact of the three new drivers of the demand—convenience, health and pleasure—is the strongest. High-performance businesses know this, and they have positioned themselves to take full advantage. Accenture defines a high-performance business as one that consistently outperforms its peers through economic and industry cycles as well as changes in leadership. Our extensive cross-industry research has identified three building blocks that underpin high performance in any industry—market focus and position; distinctive capabilities; and performance anatomy—and that all high-performance businesses have mastered. Here is what these building blocks look like in the food and nonalcoholic beverages industry.

Chapter 4

Biochemistry of Food

Nutrition (also called nourishment or aliment) is the provision, to cells and organisms, of the materials necessary (in the form of food) to support life. Many common health problems can be prevented or alleviated with a healthy diet. The diet of an organism is what it eats, which is largely determined by the perceived palatability of foods. Dietitians are health professionals who specialize in human nutrition, meal planning, economics, and preparation. They are trained to provide safe, evidence-based dietary advice and management to individuals (in health and disease), as well as to institutions.

Overview

Nutrition science investigates the metabolic and physiological responses of the body to diet. With advances in the fields of molecular biology, biochemistry, and genetics, the study of nutrition is increasingly concerned with metabolism and metabolic pathways: the sequences of biochemical steps through which substances in living things change from one form to another.

Nitrogen is needed by animals to build proteins. Carnivore and herbivore diets vary in their source of nitrogen, which is a limiting nutrient for both. Herbivores consume plants to get nitrogen and carnivores consume other animals to obtain nitrogen. Nitrogen is a common element in the atmosphere but exists in a state that is not usable by most living organisms. Certain fungi and bacteria are able to convert atmospheric nitrogen into a form plants can adsorb and utilize. The human body contains chemical compounds, such as water, carbohydrates (sugar, starch, and fiber), amino acids (in proteins), fatty acids (in lipids), and nucleic acids (DNA and RNA).

These compounds in turn consist of elements such as carbon, hydrogen, oxygen, nitrogen, phosphorus, calcium, iron, zinc, magnesium, manganese, and so on. All of these chemical compounds and elements occur in various forms and combinations (e.g. hormones, vitamins, phospholipids, hydroxyapatite), both in the human body and in the plant and animal organisms that humans eat.

The human body consists of elements and compounds ingested, digested, absorbed, and circulated through the bloodstream to feed the cells of the body. Except in the unborn fetus, which receive processed nutrients from the mother, the digestive system is the first system involved in breaking down food prior to further digestion.

Digestive juices, excreted into the lumen of the gastrointestinal tract, break chemical bonds in ingested molecules, and modulate their conformations and energy states. Though some molecules are absorbed into the bloodstream unchanged, digestive processes release them from the matrix of foods. Unabsorbed matter, along with some waste products of metabolism, is eliminated from the body in the feces.

Studies of nutritional status must take into account the state of the body before and after experiments, as well as the chemical composition of the whole diet and of all material excreted and eliminated from the body (in urine and feces). Comparing the food to the waste can help determine the specific compounds and elements absorbed and metabolized in the body.

The effects of nutrients may only be discernible over an extended period, during which all food and waste must be analysed. The number of variables involved in such experiments is high, making nutritional studies time-consuming and expensive, which explains why the science of human nutrition is still slowly evolving.

In general, eating a wide variety of fresh, whole (unprocessed), foods has proven favourable for one's health compared to monotonous diets based on processed foods. In particular, the consumption of whole-plant foods slows digestion and allows better absorption, and a more favourable balance of essential nutrients per Calorie, resulting in better management of cell growth, maintenance, and mitosis (cell division), as well as better regulation of appetite and blood sugar. Regularly scheduled meals (every few hours) have also proven more wholesome than infrequent or haphazard ones, although a recent study has also linked more frequent meals with a higher risk of colon cancer in men.

Nutrients

There are six major classes of nutrients: carbohydrates, fats, minerals, protein, vitamins, and water. These nutrient classes can be categorized as either macronutrients (needed in relatively large amounts) or micronutrients (needed in smaller quantities). The macronutrients include carbohydrates, fats, protein, and water. The micronutrients are minerals and vitamins.

The macronutrients (excluding water) provide structural material (amino acids from which proteins are built, and lipids from which cell membranes and some signalling molecules are built), and energy.

Some of the structural material can be used to generate energy internally, and in either case it is measured in Joules or kilocalories (often called "Calories" and written with a capital *C* to distinguish them from little 'c' calories). Carbohydrates and proteins provide 17 kJ approximately (4 kcal) of energy per gram, while fats provide 37 kJ (9 kcal) per gram., though the net energy from either depends on such factors as absorption and digestive effort, which vary substantially from instance to instance. Vitamins, minerals, fiber, and water provide energy, but are required for other reasons. A third class of dietary material, fiber (i.e., non-digestible material such as cellulose), is also required, for both mechanical and biochemical reasons, although the exact reasons remain unclear.

Molecules of carbohydrates and fats consist of carbon, hydrogen, and oxygen atoms. Carbohydrates range from simple monosaccharides (glucose, fructose, galactose) to complex polysaccharides (starch). Fats are triglycerides, made of assorted fatty acid monomers bound to glycerol backbone. Some fatty acids, but not all, are essential in the diet: they cannot be synthesized in the body. Protein molecules contain nitrogen atoms in addition to carbon, oxygen, and hydrogen.

The fundamental components of protein are nitrogen-containing amino acids, some of which are essential in the sense that humans cannot make them internally. Some of the amino acids are convertible (with the expenditure of energy) to glucose and can be used for energy production just as ordinary glucose in a process known as gluconeogenesis. By breaking down existing protein, some glucose can be produced internally; the remaining amino acids are discarded, primarily as urea in urine. This occurs normally only during prolonged starvation.

Other micronutrients include antioxidants and phytochemicals, which are said to influence (or protect) some body systems. Their necessity is not as well established as in the case of, for instance, vitamins. Most foods contain a mix of some or all of the nutrient classes, together with other substances, such as toxins of various sorts. Some nutrients can be stored internally (e.g., the fat soluble vitamins), while others are required more or less continuously. Poor health can be caused by a lack of required nutrients or, in extreme cases, too much of a required nutrient. For example, both salt and water (both absolutely required) will cause illness or even death in excessive amounts.

Carbohydrates

Carbohydrates include sugars, starches and fiber. They constitute a large part of foods such as rice, noodles, bread, and other grain-based products. Carbohydrates may be classified chemically as monosaccharides, disaccharides, or polysaccharides depending on the number of monomer (saccharide or sugar) units they contain. Monosaccharides, disaccharides, and polysaccharides contain one, two, and three or more sugar units, respectively.

Polysaccharides are often referred to as *complex* carbohydrates because they consist of long, sometimes branched chains of single sugar units. Mono-and disaccharides are called simple carbohydrates. Dietary advice frequently but erroneously suggests that complex carbohydrates are superior to simple because they take longer to digest and absorb. Simple carbohydrates, on the other hand, are said to cause a spike in blood glucose levels rapidly after ingestion.

These traditional claims are false. In fact, many digestible polysaccharides are processed as rapidly as simple sugars in the human body. On the other hand some simple carbohydrates (fructose, for example) are processed in a different way and do not spike blood sugar. Thus the distinction between "complex" and "simple" does not predict the nutritional value or impact of carbohydrates. A better way of determining what effect particular foods may have on blood sugar and ultimately on health in general is the glycemic index.

Carbohydrates are not essential nutrients (with the likely exception of fiber), but are typically an important part of the human diet. While it would not be accurate to categorize all carbohydrates as "bad" nutritionally, some carbohydrate sources may well have deleterious effects on health, especially when consumed in large quantities. Highly processed carbohydrates (sugars and starches) as well as fructose consumed in large quantities have been implicated in negative health outcomes.

Fiber

Dietary fiber is a carbohydrate (or a polysaccharide) that is incompletely absorbed in humans and in some animals. Like all carbohydrates, when it is metabolized it can produce four Calories (kilocalories) of energy per gram. However, in most circumstances it accounts for less than that because of its limited absorption and digestibility. Dietary fiber consists mainly of cellulose, a large carbohydrate polymer that is indigestible because humans do not have the required enzymes to disassemble it. There are two subcategories: soluble and insoluble fiber. Whole grains, fruits (especially plums, prunes, and figs), and vegetables are good sources of dietary fiber. There are many health benefits of a high-fiber diet. Dietary fiber helps reduce the chance of gastrointestinal problems such as constipation and diarrhea by increasing the weight and size of stool and softening it. Insoluble fiber, found in whole-wheat flour, nuts and vegetables, especially stimulates peristalsis — the rhythmic muscular contractions of the intestines which move digesta along the digestive tract. Soluble fiber, found in oats, peas, beans, and many fruits, dissolves in water in the intestinal tract to produce a gel which slows the movement of food through the intestines. This may help lower blood glucose levels because it can slow the absorption of sugar. Additionally, fiber, perhaps especially that from whole grains, is thought to possibly help lessen insulin spikes, and therefore reduce the risk of type 2 diabetes. The link between increased fiber consumption and a decreased risk of colorectal cancer is still uncertain.

Fat

A molecule of dietary fat typically consists of several fatty acids (containing long chains of carbon and hydrogen atoms), bonded to a glycerol. They are typically found as triglycerides (three fatty acids attached to one glycerol backbone). Fats may be classified as saturated or unsaturated depending on the detailed structure of the fatty acids involved. Saturated fats have all of the carbon atoms in their fatty acid chains bonded to hydrogen atoms, whereas unsaturated fats have some of these carbon atoms double-bonded, so their molecules have relatively fewer hydrogen atoms than a saturated fatty acid of the same length.

Unsaturated fats may be further classified as monounsaturated (one double-bond) or polyunsaturated (many double-bonds). Furthermore, depending on the location of the double-bond in the fatty acid chain, unsaturated fatty acids are classified as omega-3 or omega-6 fatty acids. Trans fats are a type of unsaturated fat with *trans-*

isomer bonds; these are rare in nature and in foods from natural sources; they are typically created in an industrial process called (partial) hydrogenation. There are nine kilocalories in each gram of fat.

Saturated fats (typically from animal sources) have been a staple in many world cultures for millennia. Unsaturated fats (e.g., vegetable oil) are considered healthier (citation needed), while trans fats are to be avoided. Saturated and some trans fats are typically solid at room temperature (such as butter or lard), while unsaturated fats are typically liquids (such as olive oil or flaxseed oil). Trans fats are very rare in nature, and have been shown to be highly detrimental to human health, but have properties useful in the food processing industry, such as rancidity resistance.

Essential Fatty Acids

Most fatty acids are non-essential, meaning the body can produce them as needed, generally from other fatty acids and always by expending energy to do so. However, in humans, at least two fatty acids are essential and must be included in the diet. An appropriate balance of essential fatty acids—omega-3 and omega-6 fatty acids—seems also important for health, although definitive experimental demonstration has been elusive. Both of these "omega" long-chain polyunsaturated fatty acids are substrates for a class of eicosanoids known as prostaglandins, which have roles throughout the human body. They are hormones, in some respects.

The omega-3 eicosapentaenoic acid (EPA), which can be made in the human body from the omega-3 essential fatty acid alpha-linolenic acid (LNA), or taken in through marine food sources, serves as a building block for series 3 prostaglandins (e.g. weakly inflammatory PGE3). The omega-6 dihomo-gamma-linolenic acid (DGLA) serves as a building block for series 1 prostaglandins (e.g. anti-inflammatory PGE1), whereas arachidonic acid (AA) serves as a building block for series 2 prostaglandins (e.g. pro-inflammatory PGE 2). Both DGLA and AA can be made from the omega-6 linoleic acid (LA) in the human body, or can be taken in directly through food. An appropriately balanced intake of omega-3 and omega-6 partly determines the relative production of different prostaglandins, which is one reason why a balance between omega-3 and omega-6 is believed important for cardiovascular health. In industrialized societies, people typically consume large amounts of processed vegetable oils, which have reduced amounts of the essential fatty acids along with too much of omega-6 fatty acids relative to omega-3 fatty acids.

The conversion rate of omega-6 DGLA to AA largely determines the production of the prostaglandins PGE1 and PGE2. Omega-3 EPA prevents AA from being released from membranes, thereby skewing prostaglandin balance away from pro-inflammatory PGE2 (made from AA) toward anti-inflammatory PGE1 (made from DGLA). Moreover, the conversion (desaturation) of DGLA to AA is controlled by the enzyme delta-5-desaturase, which in turn is controlled by hormones such as insulin (up-regulation) and glucagon (down-regulation). The amount and type of carbohydrates consumed, along with some types of amino acid, can influence processes involving insulin, glucagon, and other hormones; therefore the ratio of omega-3 versus omega-6 has wide effects on general health, and specific effects on immune function and inflammation, and mitosis (i.e. cell division).

Protein

Proteins are the basis of many animal body structures (e.g. muscles, skin, and hair). They also form the enzymes that control chemical reactions throughout the body. Each molecule is composed of amino acids, which are characterized by inclusion of nitrogen and sometimes sulphur (these components are responsible for the distinctive smell of burning protein, such as the keratin in hair). The body requires amino acids to produce new proteins (protein retention) and to replace damaged proteins (maintenance). As there is no protein or amino acid storage provision, amino acids must be present in the diet. Excess amino acids are discarded, typically in the urine.

For all animals, some amino acids are *essential* (an animal cannot produce them internally) and some are *non-essential* (the animal can produce them from other nitrogen-containing compounds). Twenty-one proteinogenic amino acids are found in the human body, along with non-proteinogenic amino acids (e.g. gamma-aminobutyric acid).

Ten of the proteinogenic amino acids are essential and, therefore, must be included in the diet. A diet that contains adequate amounts of amino acids (especially those that are essential) is particularly important in some situations: during early development and maturation, pregnancy, lactation, or injury (a burn, for instance). A *complete* protein source contains all the essential amino acids; an *incomplete* protein source lacks one or more of the essential amino acids.

Sources of dietary protein include meats, tofu and other soy-products, eggs, legumes, and dairy products such as milk and cheese. Excess amino acids from protein can be converted into glucose and used for fuel through a process called gluconeogenesis. The amino acids remaining after such conversion are discarded.

Minerals

Dietary minerals are the chemical elements required by living organisms, other than the four elements carbon, hydrogen, nitrogen, and oxygen that are present in nearly all organic molecules. The term "mineral" is archaic, since the intent is to describe simply the less common elements in the diet. Some are heavier than the four just mentioned, including several metals, which often occur as ions in the body. Some dietitians recommend that these be supplied from foods in which they occur naturally, or at least as complex compounds, or sometimes even from natural inorganic sources (such as calcium carbonate from ground oyster shells). Some minerals are absorbed much more readily in the ionic forms found in such sources. On the other hand, minerals are often artificially added to the diet as supplements; the most famous is likely iodine in iodized salt which prevents goiter.

Macrominerals

Many elements are essential in relative quantity; they are usually called "bulk minerals". Some are structural, but many play a role as electrolytes. Elements with recommended dietary allowance (RDA) greater than 200 mg/day are, in alphabetical order (with informal or folk-medicine perspectives in parentheses):

- Calcium, a common electrolyte, but also needed structurally (for muscle and digestive system health, bone strength, some forms neutralize acidity, may help clear toxins, provides signalling ions for nerve and membrane functions)
- Chlorine as chloride ions; very common electrolyte;
- Magnesium, required for processing ATP and related reactions (builds bone, causes strong peristalsis, increases flexibility, increases alkalinity)
- Phosphorus, required component of bones; essential for energy processing
- Potassium, a very common electrolyte (heart and nerve health)
- Sodium, a very common electrolyte; not generally found in dietary supplements, despite being needed in large quantities, because the ion is very common in food: typically as sodium chloride, or common salt. Excessive sodium consumption can deplete calcium and magnesium, leading to high blood pressure

and osteoporosis (Note: Some sources suggest high blood pressure is due to high water retention per osmosis).

- Sulfur, for three essential amino acids and therefore many proteins (skin, hair, nails, liver, and pancreas). Sulfur is not consumed alone, but in the form of sulfur-containing amino acids

Trace Minerals

Many elements are required in trace amounts, usually because they play a catalytic role in enzymes. Some trace mineral elements (RDA < 200 mg/day) are, in alphabetical order:

- Cobalt required for biosynthesis of vitamin B_{12} family of coenzymes. Animals cannot biosynthesize B_{12}, and must obtain this cobalt-containing vitamin in the diet
- Copper required component of many redox enzymes, including cytochrome c oxidase
- Chromium required for sugar metabolism
- Fluoride required for maintenance of bone structure
- Iodine required not only for the biosynthesis of thyroxine, but probably, for other important organs as breast, stomach, salivary glands, thymus etc.; for this reason iodine is needed in larger quantities than others in this list, and sometimes classified with the macrominerals
- Iron required for many enzymes, and for hemoglobin and some other proteins
- Manganese (processing of oxygen)
- Molybdenum required for xanthine oxidase and related oxidases
- Nickel present in urease
- Selenium required for peroxidase (antioxidant proteins)
- Vanadium (Speculative: there is no established RDA for vanadium. No specific biochemical function has been identified for it in humans, although vanadium is required for some lower organisms.)
- Zinc required for several enzymes such as carboxypeptidase, liver alcohol dehydrogenase, and carbonic anhydrase

Vitamins

Some vitamins are recognized as essential nutrients, necessary in the diet for good health. Vitamin D is the exception: it can be synthesized in the skin, in the presence of UVB radiation. Certain vitamin-like compounds that are recommended in the diet, such as carnitine, are thought useful for survival and health, but these are not "essential" dietary nutrients because the human body has some capacity to produce them from other compounds.

Moreover, thousands of different phytochemicals have recently been discovered in food (particularly in fresh vegetables), which may have desirable properties including antioxidant activity; however, experimental demonstration has been suggestive but inconclusive. Other essential nutrients that are not classified as vitamins include essential amino acids, choline, essential fatty acids, and the minerals.

Vitamin deficiencies may result in disease conditions, including goitre, scurvy, osteoporosis, impaired immune system, disorders of cell metabolism, certain forms of cancer, symptoms of premature aging, and poor psychological health (including eating disorders), among many others. Excess levels of some vitamins are also dangerous to health (notably vitamin A), and for at least one vitamin, B6, toxicity begins at levels not far above the required amount. Deficient or excess levels of minerals can also have serious health consequences.

Water

It is not fully clear how much water intake is needed by healthy people, although some assert that 6–8 glasses of water daily is the minimum to maintain proper hydration. The notion that a person should consume eight glasses of water per day cannot be traced to a credible scientific source. The effect of, greater or lesser, water intake on weight loss and on constipation is also still unclear. The original water intake recommendation in 1945 by the Food and Nutrition Board of the National Research Council read: "An ordinary standard for diverse persons is 1 milliliter for each calorie of food. Most of this quantity is contained in prepared foods."

The latest dietary reference intake report by the United States National Research Council recommended, generally, (including food sources): 2.7 litres of water total for women and 3.7 litres for men. Specifically, pregnant and breastfeeding women need additional fluids to stay hydrated. According to the Institute of Medicine—who recommend that, on average, women consume 2.2 litres and men 3.0 litres—this is recommended to be 2.4 litres (approx. 9 cups) for pregnant women and 3 litres (approx. 12.5 cups) for breastfeeding women because an especially large amount of fluid is lost during nursing.

For those who have healthy kidneys, it is somewhat difficult to drink too much water, but (especially in warm humid weather and while exercising) it is dangerous to drink too little. People can drink far more water than necessary while exercising, however, putting them at risk of water intoxication, which can be fatal. In particular, large amounts of de-ionized water are dangerous.

Normally, about 20 percent of water intake comes in food, while the rest comes from drinking water and assorted beverages (caffeinated included). Water is excreted from the body in multiple forms; including urine and feces, sweating, and by water vapour in the exhaled breath.

Antioxidants

As cellular metabolism/energy production requires oxygen, potentially damaging (e.g. mutation causing) compounds known as free radicals can form. Most of these are oxidizers (i.e. acceptors of electrons) and some react very strongly. For the continued normal cellular maintenance, growth, and division, these free radicals must be sufficiently neutralized by antioxidant compounds.

Some are produced by the human body with adequate precursors (glutathione, Vitamin C), and those the body cannot produce may only be obtained in the diet via direct sources (Vitamin C in humans, Vitamin A, Vitamin K) or produced by the body from other compounds (Beta-carotene converted to Vitamin A by the body, Vitamin D synthesized from cholesterol by sunlight).

Phytochemicals (*Section Below*) and their subgroup, polyphenols, make up the majority of antioxidants; about 4,000 are known. Different antioxidants are now known to function in a cooperative network. For example, Vitamin C can reactivate free radical-containing glutathione or Vitamin E by accepting the free radical itself. Some antioxidants are more effective than others at neutralizing different free radicals.

Some cannot neutralize certain free radicals. Some cannot be present in certain areas of free radical development (Vitamin A is fat-soluble and protects fat areas, Vitamin C is water soluble and protects those areas). When interacting with a free radical, some antioxidants produce a different free radical compound that is less dangerous or more dangerous than the previous compound. Having a variety of antioxidants allows any byproducts to be safely dealt with by more efficient antioxidants in neutralizing a free radical's butterfly effect.

Although initial studies suggested that antioxidant supplements might promote health, later large clinical trials did not detect any benefit and suggested instead that excess supplementation may be harmful.

Phytochemicals

A growing area of interest is the effect upon human health of trace chemicals, collectively called phytochemicals. These nutrients are typically found in edible plants, especially colourful fruits and vegetables, but also other organisms including seafood, algae, and fungi. The effects of phytochemicals increasingly survive rigorous testing by prominent health organizations. One of the principal classes of phytochemicals are polyphenol antioxidants, chemicals that are known to provide certain health benefits to the cardiovascular system and immune system. These chemicals are known to down-regulate the formation of reactive oxygen species, key chemicals in cardiovascular disease.

Perhaps the most rigorously tested phytochemical is zeaxanthin, a yellow-pigmented carotenoid present in many yellow and orange fruits and vegetables. Repeated studies have shown a strong correlation between ingestion of zeaxanthin and the prevention and treatment of age-related macular degeneration (AMD). Less rigorous studies have proposed a correlation between zeaxanthin intake and cataracts. A second carotenoid, lutein, has also been shown to lower the risk of contracting AMD. Both compounds have been observed to collect in the retina when ingested orally, and they serve to protect the rods and cones against the destructive effects of light.

Another carotenoid, beta-cryptoxanthin, appears to protect against chronic joint inflammatory diseases, such as arthritis. While the association between serum blood levels of beta-cryptoxanthin and substantially decreased joint disease has been established, neither a convincing mechanism for such protection nor a cause-and-effect have been rigorously studied. Similarly, a red phytochemical, lycopene, has substantial credible evidence of negative association with development of prostate cancer.

As indicated above, some of the correlations between the ingestion of certain phytochemicals and the prevention of disease are, in some cases, enormous in magnitude. Yet, even when the evidence is obtained, translating it to practical dietary advice can be difficult and counter-intuitive. Lutein, for example, occurs in many yellow and orange fruits and vegetables and protects the eyes against various diseases. However, it does not protect the eye nearly as well as zeaxanthin, and the presence of lutein in the retina will prevent zeaxanthin uptake.

Additionally, evidence has shown that the lutein present in egg yolk is more readily absorbed than the lutein from vegetable sources, possibly because of fat solubility. At the most basic level, the question "should you eat eggs?" is complex to the point of dismay, including misperceptions about the health effects of cholesterol in egg yolk, and its saturated fat content.

As another example, lycopene is prevalent in tomatoes (and actually is the chemical that gives tomatoes their red colour). It is more highly concentrated, however, in processed tomato products such as commercial pasta sauce, or tomato soup, than in fresh "healthy" tomatoes. Yet, such sauces tend to have high amounts of salt, sugar, other substances a person may wish or even need to avoid.

Intestinal Bacterial Flora

It is now also known that animal intestines contain a large population of gut flora. In humans, these include species such as *Bacteroides*, *L. acidophilus*, and *E. coli*, among many others. They are essential to digestion, and are also affected by the food we eat. Bacteria in the gut perform many important functions for humans, including breaking down and aiding in the absorption of otherwise indigestible food; stimulating cell growth; repressing the growth of harmful bacteria, training the immune system to respond only to pathogens; producing vitamin B_{12}, and defending against some infectious diseases.

Advice and Guidance

Governmental Policies

In the US, dietitians are registered (RD) or licensed (LD) with the Commission for Dietetic Registration and the American Dietetic Association, and are only able to use the title "dietitian," as described by the business and professions codes of each respective state, when they have met specific educational and experiential prerequisites and passed a national registration or licensure examination, respectively.

Anyone may call themselves a nutritionist, including unqualified dietitians, as this term is unregulated. Some states, such as the State of Florida, have begun to include the title "nutritionist" in state licensure requirements.

Most governments provide guidance on nutrition, and some also impose mandatory disclosure/labelling requirements for processed food manufacturers and restaurants to assist consumers in complying with such guidance.

In the US, nutritional standards and recommendations are established jointly by the US Department of Agriculture and US Department of Health and Human Services. Dietary and physical activity guidelines from the USDA are presented in the concept of a food pyramid, which superseded the Four Food Groups. The Senate committee currently responsible for oversight of the USDA is the *Agriculture, Nutrition and Forestry Committee.* Committee hearings are often televised on C-SPAN as seen here.

The U.S. Department of Health and Human Services provides a sample week-long menu which fulfills the nutritional recommendations of the government. Canada's Food Guide is another governmental recommendation.

Teaching

Nutrition is taught in schools in many countries. In England and Wales the Personal and Social Education and Food Technology curricula include nutrition, stressing the importance of a balanced diet and teaching how to read nutrition labels on packaging. In many schools a Nutrition class will fall within the Family and Consumer Science or Health departments. In some American schools, students are required to take a certain number of FCS or Health related classes. Nutrition is offered at many schools, and if it is not a class of its own, nutrition is included in other FCS or Health classes such as: Life Skills, Independent Living, Single Survival, Freshmen Connection, Health etc. In many Nutrition classes, students learn about the food groups, the food pyramid, Daily Recommended Allowances, calories, vitamins, minerals, malnutrition, physical activity, healthy food choices and how to live a healthy life.

A 1985 US National Research Council report entitled *Nutrition Education in US Medical Schools* concluded that nutrition education in medical schools was inadequate. Only 20% of the schools surveyed taught nutrition as a separate, required course. A 2006 survey found that this number had risen to 30%.

Healthy Diets

The United Healthcare/Pacificare nutrition guideline recommends the following foods per day to enhance your health (based on a 2,000 calorie diet)

- Grains: 6 oz per day 'Make half your grains whole'
- Vegetables: 2.5 cups per day 'Vary your veggies'
- Fruits: 2 cups per day 'Focus on fruits'

- Milk: 3 cups per day 'Get your calcium-rich foods'
- Meat and beans: 5.5 oz per day 'Go lean with protein'.

A *National Geographic* cover article from November, 2005, entitled *The Secrets of Living Longer*, looks into why certain populations have high life spans. The article is a lifestyle survey of three populations, Sardinians, Okinawans, and Adventists, who generally display longevity and "suffer a fraction of the diseases that commonly kill people in other parts of the developed world, and enjoy more healthy years of life." In sum, they offer three sets of 'best practices' to emulate. The rest is up to you. In common with all three groups is to follow a low-calorie diet that consist mainly of "fruits, vegetables, and whole grains."

The *National Geographic* article noted that an NIH funded study of 34,000 Seventh-day Adventists between 1976 and 1988 "...found that the Adventists' habit of consuming beans, soy milk, tomatoes, and other fruits lowered their risk of developing certain cancers. It also suggested that eating whole grain bread, drinking five glasses of water a day, and, most surprisingly, consuming four servings of nuts a week reduced their risk of heart disease."

The French "Paradox"

The French paradox is the observation that the French suffer a relatively low incidence of coronary heart disease, despite having a diet relatively rich in saturated fats. A number of explanations have been suggested:

- Saturated fat consumption does not cause heart disease
- Reduced consumption of processed carbohydrate and other junk foods.
- Regular consumption of red wine.
- More active lifestyles involving plenty of daily exercise, especially walking; the French are much less dependent on cars than Americans are.
- Higher consumption of artificially produced trans-fats by Americans, which has been shown to have greater lipoprotein effects per gram than saturated fat.

On the other hand, statistics collected by the World Health Organization from 1990-2000 show that the incidence of heart disease in France may have been underestimated and, in fact, may be similar to that of neighbouring countries.

Sports Nutrition

Sports nutrition is the study and practice of nutrition and diet as it relates to athletic performance. It is concerned with the type and quantity of fluid and food taken by an athlete, and deals with nutrients such as vitamins, minerals, supplements and organic substances such as carbohydrates, proteins and fats. Although an important part of many sports training regimens, it is most commonly considered in strength sports (such as weight lifting and bodybuilding) and endurance sports (for example cycling, running, triathlon).

History

Ancient

From the time when the ancient Greeks and Romans started the Olympic games, the athletes had their own special regimen for great performance which included diet and nutrition. For example, Milo of Croton, the wrestler with legendary strength who won five successive Olympic Games from 532 to 516 B.C., ate "9 kilograms (20 lb) of meat, 9 kilograms (20 lb) of bread and 8.5 litres (18 US pt) of wine a day," according to Athenaeus and Pausanias (Grandjean). Alcohol was commonly drunk as an ergogenic aid to increase performance in the Olympics through the early 1900s.

Modern

The view of sports nutrition today has much evolved from the ancient Olympic gladiators' meal plan. Scientists are continually interested in learning more about this abiding subject. Research shows that the coupling of exercise and proper diet is what produces a healthy lifestyle that can maintain the "prevention/management of [chronic diseases such as] noninsulin-independent diabetes, hypertension, coronary heart disease, osteoporosis, obesity, mental health, colon cancers, stroke and back injury."

In 2008, US News reported that 65% of Americans exercised regularly by working out, playing sports, and/or other physical activities, thus the importance of proper nutrition is of great interest to athletes and exercisers for optimal performance and long term benefits.

In addition to diet, social and cultural influences, lifestyle habits, motivation and training determine successful athletic performance (Maughan). However, states Maughan, "without proper nutrition, the full potential of the athlete will not be realized, because performance will not be at its peak, training levels may not be sustained, recovery from injury will be slower, and the athlete may become more susceptible to injury and infection." Understanding sports nutrition leads to optimal athletic performance and lifetime health benefits and can thus be evaluated by the intake of certain nutrients and supplements when exercising, by learning the way the body utilizes these materials and how these practices complement future diet and exercise of the individual.

Goals

Sports nutrition has many goals to enhance performance:

First, it improves performance by improving body composition, which increases speed, quickness, mobility, and strength.

Second, it will help the speed of recovery, which will in turn create more capacity for practicing and competition as the body is becoming more fit and adjusted to the coupling of the good nutrition incorporated into the workout regimen.

Third, it will allow one to increase energy for both practice and competition, which will definitely help one's performance. Strategic diet will also increase immunity, allowing one to stay healthy and be able to continue and intensify practice and training. Most importantly, it will improve your overall health as proper health is essential to all aspects of life.

Supplements

Many athletes consider taking dietary supplements because they are looking for the "magic ingredient" to increase performance. In the extreme case of performance-enhancing supplements, athletes (particularly bodybuilders) may choose to use illegal substances such as anabolic steroids, compounds which are related to the hormone testosterone, which can quickly build mass and strength, but have many adverse effects such as high blood pressure and negative gender specific effects. Blood doping, another illegal ergogenic, was discovered in the 1940s when it was used by World War II pilots.

Dietary protein began to be consumed in the 1940s and muscle building results were found in resistance and strength training athletes. Protein intake is a part of the nutrient requirements for the regular athlete and is an important component of exercise training, because it can also aid in performance and recovery. Dietary protein intake for well-trained athletes should occur before, during and after physical activity as it is advantageous in gaining muscle mass and strength. However, if too much protein and amino acid supplements are consumed (especially by the Average Joe exerciser), it can be more harmful than beneficial; health risks include: "dehydration, gout, calcium loss, liver, and renal damage [and] gastrointestinal side effects include diarrhea, bloating, and water loss" (Lawerence). A bountiful protein diet must be paired with a healthy, well-rounded meal plan and regular resistance exercise. Yet, characteristics such as the type of exercise, intensity, duration, the carbohydrate values of diet, the individual's sex and age and also the amount of background training and training environment.

Creatine may be helpful for well-trained athletes to increase exercise performance and strength in concordance with their dietary regimen. Also, the substance glutamine, found in whey protein supplements, is the most abundant free amino acid found in the human body. For well-trained and well-nourished athletes it is considered that glutamine may have a possible role in stimulated anabolic processes such muscle glycogen and protein synthesis. Other popular supplements studies done include androstenedione, chromium, and ephedra. The findings show that there are no substantial benefits from the extra intake of these supplements, yet higher health risks and costs.

High energy supplements have shown to increase the performance of physical activity. A study done at the University of Texas saw a 4.7% increase of performance in 83% of participants after drinking Red Bull Energy Drink which was more intense than the compared placebo. The energy drink most dominantly increased the epinephrine and noreprinephrine (adrenaline and its precursor) levels and beta-endorphins in the blood than before consumption. Caffeine, carbohydrates and Vitamin B are factors that may have favoured performance increase with no change in perceived exertion.

Caffeine has been known since the 1900s and became popularly used since the 1970s when its power of masking fatigue became highly recognized. Similarly, the caffeine found in energy drinks shows an increased reaction performance and increased good feelings of energy, focus and alterness in quickness and reaction anaerobic power tests. In other words, consuming an energy drink with caffeine increases short time/rapid exercise performance (like short full-speed sprints and heavy power weight lifting.)

Post-exercise nutrition is just as important, if not more important than pre-exercise nutrition as it pertains to recovery. Traditionally, sports drinks such as Gatorade and Powerade, are consumed during and after exercise because they effectively rehydrate the body by refueling the body with minerals and electrolytes. Gatorade was founded in the 1960s, when the University of Florida, Gainesville Gators improved their performance with "Gator Aid." A drink was made of glucose and sucrose in water and helped the football players' performance. And by the 1970s, many other sports drinks of its kind had been manufactured. However, sports drinks lack protein.

New studies in 2008 have found milk, especially skim milk and chocolate milk may be the new sports drink, as milk leads to protein the synthesis which boosts net muscle protein balance. Milk naturally contains many electrolytes, nutrients and other properties that help to make it a great post-exercise beverage. When compared to plain water or sports drinks, research suggests that chocolate milk is more effective at replacing fluids lost as sweat and maintaining normal body fluid levels. Athletes drinking chocolate milk following exercise-induced dehydration had fluid levels about 2 percent higher (on initial body mass) than those using other post-exercise recovery beverages, allowing for prolonged performance, especially in repeated bouts of exercise or training..

Factors Influencing Nutritional Requirements

Differing conditions and objectives suggest the need for athletes to ensure that their nutritional approach is appropriate for their situation. Factors that may affect an athlete's nutritional needs include type of activity (aerobic vs. anaerobic), gender, weight, height, body mass index, workout or activity stage (pre-workout, intra-workout, recovery), and time of day (e.g. some nutrients are utilized by the body more effectively during sleep than while awake).

Anaerobic Exercise

During anaerobic exercise, the process of glycolysis breaks down the sugars from carbohydrates for energy without the use of oxygen. This type of exercise occurs in physical activity such as power sprints, strength resistances and quick explosive movement where the muscles are being used for power and speed, with short time energy use. After this type of exercise, there is a need to refill glycogen storage sites in the body (the long simple sugar chains in the body that store energy), although they are not likely fully depleted. To compensate for this glycogen reduction, athletes will often take in a large amount of carbohydrates in the period immediately following exercise. Typically, high glycemic index carbohydrates are preferred for their ability to rapidly raise blood glucose levels.

For the purpose of protein synthesis, protein or individual amino acids are ingested as well. Branched-chain amino acids are important since they are most responsible for protein synthesis. Often in the continuation of this anaerobic exercise, the product from this metabolic mechanism builds up in what is called lactic acid fermentation. Lactate is produced more quickly than it is being removed and it serves to regenerate NAD^+ to the cells where it's needed. During intense exercise when oxygen is not being used, a high amount of ATP is produced and pH levels fall causing acidosis or more specifically lactic acidosis. Lactic acid build up can be treated by staying well-hydrated throughout and especially after the workout, having good cool down routine and good post-workout stretching.

Protein

Protein is an important component of every cell in the body. Hair and nails are mostly made of protein. The body uses protein to build and repair tissues. In addition, protein is used to make hormones and other chemicals in the body. Protein is also an important building block of bones, muscles, cartilage, skin, and blood.

The protein requirement for each individual differs, as do opinions about whether and to what extent physically active people require more protein. The 2005 Recommended Dietary Allowances (RDA), aimed at the general healthy adult population, provide for an intake of 0.8 - 1 grams of protein per kilogram of body weight (according to the BMI formula), with the review panel stating that "no additional dietary protein is suggested for healthy adults undertaking resistance or endurance exercise". Conversely, Di Pasquale (2008), citing recent studies, recommends a minimum protein intake of 2.2 g/kg "for anyone involved in competitive or intense recreational sports who wants to maximize lean body mass but does not wish to gain weight".

Water

Water is one of the most important nutrients in the sports diet. It helps eliminate food waste products in the body, regulates body temperature during activity and helps with digestion. Maintaining hydration during periods of physical exertion is key to peak performance. While drinking too much water during activities can lead to physical discomfort, dehydration in excess of 2% of body mass (by weight) markedly hinders athletic performance. Additional carbohydrates and protein before, during, and after exercise increase time to exhaustion as well as speed recovery. The amount of water needed is based on work performed, lean body mass, and environmental factors, especially ambient temperature and humidity.

Carbohydrates

The main fuel used by the body during exercise is carbohydrates, which are stored in muscle as glycogen—a form of sugar. During exercise, muscle glycogen reserves can be used up, especially when activities last longer than 90 min. Because the amount of glycogen stored in the body is limited, it is important for athletes to replace glycogen by consuming a diet high in carbohydrates. Meeting energy needs can help improve performance during the sport, as well as improve overall strength and endurance.

There are different kinds of carbohydrates—simple or refined, and unrefined. A typical American consumes about 50% of their carbohydrates as simple sugars, which are added to foods as opposed to sugars that come naturally in fruits and vegetables.

These simple sugars come in large amounts in sodas and fast food. Over the course of a year, the average American consumes 54 gallons of soft drinks, which contain the highest amount of added sugars. Even though carbohydrates are necessary for humans to function, they are not all equally healthful. When machinery has been used to remove bits of high fiber, the carbohydrates are refined. These are the carbohydrates found in white bread and fast food.

Malnutrition

Malnutrition refers to insufficient, excessive, or imbalanced consumption of nutrients. In developed countries, the diseases of malnutrition are most often associated with nutritional imbalances or excessive consumption. Although there are more people in the world who are malnourished due to excessive consumption, according to the United Nations World Health Organization, the real challenge in developing nations today, more than starvation, is combating insufficient nutrition — the lack of nutrients necessary for the growth and maintenance of vital functions.

Mental Agility

Research indicates that improving the awareness of nutritious meal choices and establishing long-term habits of healthy eating have a positive effect on cognitive and spatial memory capacity, potentially increasing a student's potential to process and retain academic information. Some organizations have begun working with teachers, policymakers, and managed foodservice contractors to mandate improved nutritional content and increased nutritional resources in school cafeterias from primary to university level institutions. Health and nutrition have been proven to have close links with overall educational success. Currently, less than 10% of American college students report that they eat the recommended five servings of fruit and vegetables daily. Better nutrition has been shown to have an impact on both cognitive and spatial memory performance; a study showed those with higher blood sugar levels performed better on certain memory tests. In another study, those who consumed yogurt performed better on thinking tasks when compared to those who consumed caffeine free diet soda or confections. Nutritional deficiencies have been shown to have a negative effect on learning behaviour in mice as far back as 1951.

"Better learning performance is associated with diet induced effects on learning and memory ability". The "nutrition-learning nexus" demonstrates the correlation between diet and learning and has application in a higher education setting.

"We find that better nourished children perform significantly better in school, partly because they enter school earlier and thus have more time to learn but mostly because of greater learning productivity per year of schooling."

91% of college students feel that they are in good health while only 7% eat their recommended daily allowance of fruits and vegetables.

Nutritional education is an effective and workable model in a higher education setting. More "engaged" learning models that encompass nutrition is an idea that is picking up steam at all levels of the learning cycle.

There is limited research available that directly links a student's Grade Point Average (G.P.A.) to their overall nutritional health. Additional substantive data is needed to prove that overall intellectual health is closely linked to a person's diet, rather than just another correlation fallacy.

Mental Disorders

Nutritional supplement treatment may be appropriate for major depression, bipolar disorder, schizophrenia, and obsessive compulsive disorder, the four most common mental disorders in developed countries. Supplements that have been studied most for mood elevation and stabilization include eicosapentaenoic acid and docosahexaenoic acid (each of which are an omega-3 fatty acid contained in fish oil, but not in flaxseed oil), vitamin B_{12}, folic acid, and inositol.

Cancer

Cancer is now common in developing countries. According to a study by the International Agency for Research on Cancer, "In the developing world, cancers of the liver, stomach and esophagus were more common, often linked to consumption of carcinogenic preserved foods, such as smoked or salted food, and parasitic infections that attack organs." Lung cancer rates are rising rapidly in poorer nations because of increased use of tobacco. Developed countries "tended to have cancers linked to affluence or a 'Western lifestyle' — cancers of the colon, rectum, breast and prostate — that can be caused by obesity, lack of exercise, diet and age."

Metabolic Syndrome

Several lines of evidence indicate lifestyle-induced hyperinsulinemia and reduced insulin function (i.e. insulin resistance) as a decisive factor in many disease states. For example, hyperinsulinemia and insulin resistance are strongly linked to chronic inflammation, which in turn is strongly linked to a variety of adverse developments such as arterial microinjuries and clot formation (i.e. heart disease) and exaggerated cell division (i.e. cancer). Hyperinsulinemia and insulin resistance (the so-called metabolic syndrome) are characterized by a combination of abdominal obesity, elevated blood sugar, elevated blood pressure, elevated blood triglycerides, and reduced HDL cholesterol. The negative impact of hyperinsulinemia on prostaglandin PGE1/PGE2 balance may be significant.

The state of obesity clearly contributes to insulin resistance, which in turn can cause type 2 diabetes. Virtually all obese and most type 2 diabetic individuals have marked insulin resistance. Although the association between overweight and insulin resistance is clear, the exact (likely multifarious) causes of insulin resistance remain less clear. Importantly, it has been demonstrated that appropriate exercise, more regular food intake and reducing glycemic load all can reverse insulin resistance in overweight individuals (and thereby lower blood sugar levels in those who have type 2 diabetes).

Obesity can unfavourably alter hormonal and metabolic status via resistance to the hormone leptin, and a vicious cycle may occur in which insulin/leptin resistance and obesity aggravate one another.

The vicious cycle is putatively fuelled by continuously high insulin/leptin stimulation and fat storage, as a result of high intake of strongly insulin/leptin stimulating foods and energy. Both insulin and leptin normally function as satiety signals to the hypothalamus in the brain; however, insulin/leptin resistance may reduce this signal and therefore allow continued overfeeding despite large body fat stores. In addition, reduced leptin signalling to the brain may reduce leptin's normal effect to maintain an appropriately high metabolic rate.

There is a debate about how and to what extent different dietary factors— such as intake of processed carbohydrates, total protein, fat, and carbohydrate intake, intake of saturated and trans fatty acids, and low intake of vitamins/minerals—contribute to the development of insulin and leptin resistance. In any case, analogous to the way modern man-made pollution may potentially overwhelm the environment's ability to maintain homeostasis, the recent explosive introduction of high glycemic index and processed foods into the human diet may potentially overwhelm the body's ability to maintain homeostasis and health (as evidenced by the metabolic syndrome epidemic).

Hyponatremia

Excess water intake, without replenishment of sodium and potassium salts, leads to hyponatremia, which can further lead to water intoxication at more dangerous levels. A well-publicized case occurred in 2007, when Jennifer Strange died while participating in a water-drinking contest. More usually, the condition occurs in long-distance endurance events (such as marathon or triathlon competition and training) and causes gradual mental dulling, headache, drowsiness, weakness, and confusion; extreme cases may result in coma, convulsions, and death. The primary damage comes from swelling of the brain, caused by increased osmosis as blood salinity decreases.

Antinutrient

Antinutrients are natural or synthetic compounds that interfere with the absorption of nutrients. Nutrition studies focus on antinutrients commonly found in food sources and beverages.

Processed Foods

Since the Industrial Revolution some two hundred years ago, the food processing industry has invented many technologies that both help keep foods fresh longer and alter the fresh state of food as they appear in nature. Cooling is the primary technology used to maintain freshness, whereas many more technologies have been invented to allow foods to last longer without becoming spoiled. These latter technologies include pasteurisation, autoclavation, drying, salting, and separation of various components, all of which appear to alter the original nutritional contents of food. Pasteurisation and autoclavation (heating techniques) have no doubt improved the safety of many common foods, preventing epidemics of bacterial infection. But some of the (new) food processing technologies undoubtedly have downfalls as well.

Modern separation techniques such as milling, centrifugation, and pressing have enabled concentration of particular components of food, yielding flour, oils, juices and so on, and even separate fatty acids, amino acids, vitamins, and minerals. Inevitably, such large scale concentration changes the nutritional content of food, saving certain nutrients while removing others. Heating techniques may also reduce food's content of many heat-labile nutrients such as certain vitamins and phytochemicals, and possibly other yet to be discovered substances.

Because of reduced nutritional value, processed foods are often 'enriched' or 'fortified' with some of the most critical nutrients (usually certain vitamins) that were lost during processing. Nonetheless, processed foods tend to have an inferior nutritional profile compared to whole, fresh foods, regarding content of both sugar and high GI starches, potassium/sodium, vitamins, fiber, and of intact, unoxidized (essential) fatty acids. In addition, processed foods often contain potentially harmful substances such as oxidized fats and trans fatty acids.

A dramatic example of the effect of food processing on a population's health is the history of epidemics of beri-beri in people subsisting on polished rice. Removing the outer layer of rice by polishing it removes with it the essential vitamin thiamine, causing beri-beri. Another example is the development of scurvy among infants in the late 19th century in the United States. It turned out that the vast majority of sufferers were being fed milk that had been heat-treated (as suggested by Pasteur) to control bacterial disease. Pasteurisation was effective against bacteria, but it destroyed the vitamin C.

As mentioned, lifestyle-and obesity-related diseases are becoming increasingly prevalent all around the world. There is little doubt that the increasingly widespread application of some modern food processing technologies has contributed to this development.

The food processing industry is a major part of modern economy, and as such it is influential in political decisions (e.g. nutritional recommendations, agricultural subsidising). In any known profit-driven economy, health considerations are hardly a priority; effective production of cheap foods with a long shelf-life is more the trend. In general, whole, fresh foods have a relatively short shelf-life and are less profitable to produce and sell than are more processed foods.

Thus, the consumer is left with the choice between more expensive, but nutritionally superior, whole, fresh foods, and cheap, usually nutritionally inferior, processed foods. Because processed foods are often cheaper, more convenient (in both purchasing, storage, and preparation), and more available, the consumption of nutritionally inferior foods has been increasing throughout the world along with many nutrition-related health complications.

History

Humans have evolved as omnivorous hunter-gatherers over the past 250,000 years. The diet of early modern humans varied significantly depending on location and climate. The diet in the tropics tended to be based more heavily on plant foods, while the diet at higher latitudes tended more towards animal products. Analysis of postcranial and cranial remains of humans and animals from the Neolithic, along with detailed bone modification studies have shown that cannibalism was also prevalent among prehistoric humans.

Agriculture developed about 10,000 years ago in multiple locations throughout the world, providing grains such as wheat, rice, potatoes, and maize, with staples such as bread, pasta, and tortillas. Farming also provided milk and dairy products, and sharply increased the availability of meats and the diversity of vegetables. The importance of food purity was recognized when bulk storage led to infestation and contamination risks. Cooking developed as an often ritualistic activity, due to efficiency and reliability concerns requiring adherence to strict recipes and procedures, and in response to demands for food purity and consistency.

From Antiquity to 1900

The first recorded nutritional experiment is found in the Bible's Book of Daniel. Daniel and his friends were captured by the king of Babylon during an invasion of Israel. Selected as court servants, they were to share in the king's fine foods and wine. But they objected, preferring vegetables (pulses) and water in accordance with their

Jewish dietary restrictions. The king's chief steward reluctantly agreed to a trial. Daniel and his friends received their diet for 10 days and were then compared to the king's men. Appearing healthier, they were allowed to continue with their diet.

Around 475 BC, Anaxagoras stated that food is absorbed by the human body and therefore contained "homeomerics" (generative components), suggesting the existence of nutrients. Around 400 BC, Hippocrates said, "Let food be your medicine and medicine be your food."

In the 16th century, scientist and artist Leonardo da Vinci compared metabolism to a burning candle. In 1747, Dr. James Lind, a physician in the British navy, performed the first scientific nutrition experiment, discovering that lime juice saved sailors who had been at sea for years from scurvy, a deadly and painful bleeding disorder. The discovery was ignored for forty years, after which British sailors became known as "limeys." The essential vitamin C within lime juice would not be identified by scientists until the 1930s.

Around 1770, Antoine Lavoisier, the "Father of Nutrition and Chemistry" discovered the details of metabolism, demonstrating that the oxidation of food is the source of body heat. In 1790, George Fordyce recognized calcium as necessary for fowl survival. In the early 19th century, the elements carbon, nitrogen, hydrogen and oxygen were recognized as the primary components of food, and methods to measure their proportions were developed.

In 1816, François Magendie discovered that dogs fed only carbohydrates and fat lost their body protein and died in a few weeks, but dogs also fed protein survived, identifying protein as an essential dietary component. In 1840, Justus Liebig discovered the chemical makeup of carbohydrates (sugars), fats (fatty acids) and proteins (amino acids.) In the 1860s, Claude Bernard discovered that body fat can be synthesized from carbohydrate and protein, showing that the energy in blood glucose can be stored as fat or as glycogen.

In the early 1880s, Kanehiro Takaki observed that Japanese sailors (whose diets consisted almost entirely of white rice) developed beriberi (or endemic neuritis, a disease causing heart problems and paralysis), but British sailors and Japanese naval officers did not. Adding various types of vegetables and meats to the diets of Japanese sailors prevented the disease. In 1896, Eugen Baumann observed iodine in thyroid glands.

In 1897, Christiaan Eijkman worked with natives of Java, who also suffered from beriberi. Eijkman observed that chickens fed the native diet of white rice developed the symptoms of beriberi, but remained healthy when fed unprocessed brown rice with the outer bran intact. Eijkman cured the natives by feeding them brown rice, discovering that food can cure disease. Over two decades later, nutritionists learned that the outer rice bran contains vitamin B1, also known as thiamine.

From 1900 to the Present

In the early 20th century, Carl Von Voit and Max Rubner independently measured caloric energy expenditure in different species of animals, applying principles of physics in nutrition. In 1906, Wilcock and Hopkins showed that the amino acid tryptophan was necessary for the survival of rats. He fed them a special mixture of food containing all the nutrients he believed were essential for survival, but the rats died. A second group of rats were fed an amount of milk containing vitamins. Gowland Hopkins recognized "accessory food factors" other than calories, protein and minerals, as organic materials essential to health, but which the body cannot synthesize. In 1907, Stephen M. Babcock and Edwin B. Hart conducted the single-grain experiment, which took nearly four years to complete.

In 1912, Casimir Funk coined the term vitamin, a vital factor in the diet, from the words "vital" and "amine," because these unknown substances preventing scurvy, beriberi, and pellagra, were thought then to be derived from ammonia. The vitamins were studied in the first half of the 20th century.

In 1913, Elmer McCollum discovered the first vitamins, fat soluble vitamin A, and water soluble vitamin B (in 1915; now known to be a complex of several water-soluble vitamins) and named vitamin C as the then-unknown substance preventing scurvy.

Lafayette Mendel and Thomas Osborne also performed pioneering work on vitamins A and B. In 1919, Sir Edward Mellanby incorrectly identified rickets as a vitamin A deficiency because he could cure it in dogs with cod liver oil. In 1922, Elmer McCollum destroyed the vitamin A in cod liver oil, but found that it still cured rickets. Also in 1922, H.M. Evans and L.S. Bishop discover vitamin E as essential for rat pregnancy, originally calling it "food factor X" until 1925.

In 1925, Hart discovered that trace amounts of copper are necessary for iron absorption. In 1927, Adolf Otto Reinhold Windaus synthesized vitamin D, for which he won the Nobel Prize in Chemistry in 1928. In 1928, Albert Szent-Györgyi isolated ascorbic acid, and in 1932 proved that it is vitamin C by preventing scurvy. In 1935 he synthesized it, and in 1937, he won a Nobel Prize for his efforts. Szent-Györgyi concurrently elucidated much of the citric acid cycle.

In the 1930s, William Cumming Rose identified essential amino acids, necessary protein components which the body cannot synthesize. In 1935, Underwood and Marston independently discovered the necessity of cobalt. In 1936, Eugene Floyd Dubois showed that work and school performance are related to caloric intake. In 1938, Erhard Fernholz discovered the chemical structure of vitamin E. It was synthesised by Paul Karrer.

In 1940, rationing in the United Kingdom during and after World War II took place according to nutritional principles drawn up by Elsie Widdowson and others. In 1941, the first Recommended Dietary Allowances (RDAs) were established by the National Research Council.

In 1992, The U.S. Department of Agriculture introduced the Food Guide Pyramid. In 2002, a Natural Justice study showed a relation between nutrition and violent behaviour. In 2005, a study found that obesity may be caused by adenovirus in addition to bad nutrition.

Plant Nutrition

Plant nutrition is the study of the chemical elements that are necessary for plant growth. There are several principles that apply to plant nutrition. Some elements are directly involved in plant metabolism. However, this principle does not account for the so-called beneficial elements, whose presence, while not required, has clear positive effects on plant growth. A nutrient that is able to limit plant growth according to Liebig's law of the minimum, is considered an essential plant nutrient if the plant cannot complete its full life cycle without it. There are 17 essential plant nutrients.

Macronutrients:

- N = Nitrogen
- P = Phosphorus
- K = Potassium
- Ca = Calcium
- Mg = Magnesium
- S = Sulfur
- Si = Silicon.

Micronutrients (trace levels) include:

- Cl = Chlorine
- Fe = Iron
- B = Boron
- Mn = Manganese
- Na = Sodium
- Zn = Zinc
- Cu = Copper
- Ni= Nickel
- Mo = Molybdenum.

Macronutrients

Calcium

Calcium regulates transport of other nutrients into the plant and is also involved in the activation of certain plant enzymes. Calcium deficiency results in stunting.

Nitrogen

Nitrogen is an essential component of all proteins. Nitrogen deficiency most often results in stunted growth.

Phosphorus

Phosphorus is important in plant bioenergetics. As a component of ATP, phosphorus is needed for the conversion of light energy to chemical energy (ATP) during photosynthesis. Phosphorus can also be used to modify the activity of various enzymes by phosphorylation, and can be used for cell signalling. Since ATP can be used for the biosynthesis of many plant biomolecules, phosphorus is important for plant growth and flower/seed formation.

Potassium

Potassium regulates the opening and closing of the stoma by a potassium ion pump. Since stomata are important in water regulation, potassium reduces water loss from the leaves and increases drought tolerance. Potassium deficiency may cause necrosis or interveinal chlorosis.

Silicon

Silicon is deposited in cell walls and contributes to its mechanical properties including rigidity and elasticity

Micronutrients

Boron

Boron is important in sugar transport, cell division, and synthesizing certain enzymes. Boron deficiency causes necrosis in young leaves and stunting.

Copper

Copper is important for photosynthesis. Symptoms for copper deficiency include chlorosis. Involved in many enzyme processes. Necessary for proper photosythesis. Involved in the manufacture of lignin (cell walls). Involved in grain production.

Chlorine

Chlorine is necessary for osmosis and ionic balance; it also plays a role in photosynthesis.

Iron

Iron is necessary for photosynthesis and is present as an enzyme cofactor in plants. Iron deficiency can result in interveinal chlorosis and necrosis.

Manganese

Manganese is necessary for building the chloroplasts. Manganese deficiency may result in coloration abnormalities, such as discolored spots on the foliage.

Molybdenum

Molybdenum is a cofactor to enzymes important in building amino acids.

Nickel

In higher plants, Nickel is essential for activation of urease, an enzyme involved with nitrogen metabolism that is required to process urea. Without Nickel, toxic levels of urea accumulate, leading to the formation of necrotic lesions. In lower plants, Nickel activates several enzymes involved in a variety of processes, and can substitute for Zinc and Iron as a cofactor in some enzymes.

Sodium

Sodium is involved in the regeneration of phosphoenolpyruvate in CAM and C4 plants. It can also substitute for potassium in some circumstances.

Zinc

Zinc is required in a large number of enzymes and plays an essential role in DNA transcription. A typical symptom of zinc deficiency is the stunted growth of leaves, commonly known as "little leaf" and is caused by the oxidative degradation of the growth hormone auxin.

Processes

Plants uptake essential elements from the soil through their roots and from the air (mainly consisting of nitrogen and oxygen) through their leaves. Nutrient uptake in the soil is achieved by cation exchange, wherein root hairs pump hydrogen ions (H^+) into the soil through proton pumps. These hydrogen ions displace cations attached to negatively charged soil particles so that the cations are available for uptake by the root. In the leaves, stomata open to take in carbon dioxide and expel oxygen.

The carbon dioxide molecules are used as the carbon source in photosynthesis. Although nitrogen is plentiful in the Earth's atmosphere, relatively few plants engage in nitrogen fixation (conversion of atmospheric nitrogen to a biologically useful form). Most plants therefore require nitrogen compounds to be present in the soil in which they grow.

Plant nutrition is a difficult subject to understand completely, partially because of the variation between different plants and even between different species or individuals of a given clone.

Elements present at low levels may cause deficiency symptoms, and toxicity is possible at levels that are too high. Furthermore, deficiency of one element may present as symptoms of toxicity from another element, and vice-versa.

Carbon and oxygen are absorbed from the air, while other nutrients are absorbed from the soil. Green plants obtain their carbohydrate supply from the carbon dioxide in the air by the process of photosynthesis.

Malnutrition

Malnutrition refers to insufficient, excessive, or imbalanced consumption of nutrients by an organism. In developed countries, the diseases of malnutrition are most often associated with nutritional imbalances or excessive consumption.

Although there are more organisms in the world who are malnourished due to insufficient consumption, increasingly more organisms suffer from excessive over-nutrition; a problem caused by an over abundance of sustenance coupled with the instinctual desire (by animals in particular) to consume all that it can.

Insufficient

Under consumption generally refers to the long-term consumption of insufficient sustenance in relation to the energy that an organism expends or expels, leading to poor health.

Excessive

Over consumption generally refers to the long-term consumption of excess sustenance in relation to the energy that an organism expends or expels, leading to poor health and, in animals, obesity. It can cause excessive hair loss, brittle nails, and irregular premenstrual cycles for females

Unbalanced

When too much of one or more nutrients is present in the diet to the exclusion of the proper amount of other nutrients, the diet is said to be unbalanced.

Nutritional biochemistry is the study of nutrition as a science. While nutritional science is composed of various studies in food components, nutrients, and their function regarding humans and other mammals, nutritional biochemistry specifically focuses on nutrient chemical components, and how they function metabolically, physiologically, and biochemically.

Biochemistry research in this field is primarily centered upon defining dietary regulations for the general public.

An integrative form of science, nutritional biochemistry uses other sciences, such as chemistry, biology, and physics. It applies these sciences specifically to the study of health, diet, disease, and the connections that exist among them. To do this, biochemists might ascertain the structures of different nutrients, or use a biochemistry lab to discover how specific nutrients might affect the metabolism. Such research can lead to the development of public health policies and interventions.

Malnutrition is typically a serious concern for a nutritional biochemist. It is often the cause of disabilities, disease, and other problems due to nutritional deficiencies or overnutrition.

These conditions can include diabetes, cancer, birth defects, obesity, and cardiovascular disease. Through the study of nutritional biochemistry, scientists can develop plans to counteract these afflictions, identifying both nutritional causes as well as preventative measures.

Genetics are also commonly studied by nutritional biochemists. Many diseases and dietary factors are affected by genetics; biochemists can use genetic studies to pinpoint these different effects, and improve public health overall.

The fortification of many different food products, promotion of low-cholesterol diets, and iodinization of table salt were all introduced into society through recommendations made through nutritional biochemistry findings.

Chapter 5

Food Technology

Food technology, or food tech for short is the application of food science to the selection, preservation, processing, packaging, distribution, and use of safe, nutritious, and wholesome food. Food scientists and food technologists study the physical, microbiological, and chemical make-up of food. Depending on their area of specialization, food scientists may develop ways to process, preserve, package, or store food, according to industry and government specifications and regulations. Consumers seldom think of the vast array of foods and the research and development that has resulted in the means to deliver tasty, nutritious, safe, and convenient foods., food technology is part of the curriculum and teaches, alongside *how* to cook, nutrition and the food manufacturing process.

Early History of Food Technology

Research in the field now known as food technology has been conducted for decades. Nicolas Appert's development in 1810 of the canning process was a decisive event. The process wasn't called canning then and Appert did not really know the principle on which his process worked, but canning has had a major impact on food preservation techniques. Louis Pasteur's research on the spoilage of wine and his description of how to avoid spoilage in 1864 was an early attempt to put food technology on a scientific basis. Besides research into wine spoilage, Pasteur did research on the production of alcohol, vinegar, wines and beer, and the souring of milk. He developed pasteurization—the process of heating milk and milk products to destroy food spoilage and disease-producing organisms. In his research into food technology, Pasteur became the pioneer into bacteriology and of modern preventive medicine. By 1940's to 1950's, the original four departments that had taught the subject under different names (including those at the University of Massachusetts and the University of California) had been retitled “food science”, “food science and technology”, or a similar variant.

Developments in Food Technology

Several companies in the food industry have played a role in the development of food technology. These developments have contributed greatly to the food supply. Some of these developments are:

- Instantized Milk Powder - D.D. Peebles (U.S. patent 2,835,586) developed the first instant milk powder, which has become the basis for a variety of new products that are rehydratable in cold water or milk. This process increases the surface area of the powdered product by partially rehydrating spray-dried milk powder.
- Freeze Drying - The first application of freeze drying was most likely in the pharmaceutical industry; however, a successful large-scale industrial application of the process was the development of continuous freeze drying of coffee.
- High-Temperature Short Time Processing - These processes for the most part are characterized by rapid heating and cooling, holding for a short time at a relatively high temperature and filling aseptically into sterile containers.
- Decaffeination of Coffee and Tea - Decaffeinated coffee and tea was first developed on a commercial basis in Europe around 1900. The process is described in U.S. patent 897,763. Green coffee beans are treated with steam or water to around 20% moisture.

  The added water and heat separate the caffeine from the bean to its surface. Solvents are then used to remove the caffeine from the beans. In the 1980s, new non-organic solvent techniques have been developed for the decaffeination of coffee and tea. Carbon dioxide under supercritical conditions is one of these new techniques. U.S. patent 4,820,537 was issued to General Foods Corp. for a CO_2 decaffeination process.
- Process optimization- Food Technology now allows production of foods to be more efficient, Oil saving technologies are now available on different forms. Production methods and methodology have also become increasingly sophisticated.

Technological Advancements in Food Products

Food Processing Industry Advances

America's current food-processing industry can be traced back to a tasking by Napoleon that led his quartermaster to engage the services of Nicolas Appert to produce stable rations for his troops to carry into battle. The canning industry that originated in 1809 has evolved during two centuries into a mature industry that still relies heavily on thermal processing to feed a growing population.

The Army used thermal processing to develop the C Ration, which evolved into the Meal, Combat Individual, that served U.S. troops from World War II through the Vietnam era. These rations relied on cylindrical metal cans.

In the late 1970s, the Army replaced metal cans with the more flexible and versatile polymeric retort pouch for improved thermally-processed products. The retort pouch was developed by the Natick Soldier Center, part of the U.S. Army Soldier Systems Center (Natick), in cooperation with industry and academia, and it is the central feature of the Meal, Ready-to-Eat (MRE). The 21st version of this ration, which will be procured in 2001, will have 24 different meals, each with a different entree in a retort pouch.

To increase the mobility of individual soldiers, lightweight, calorically-dense or dehydrated rations are required. The freeze-dehydration process pioneered by the Natick Soldier Center is capable of producing high-quality meals with long shelf life. This process produced the Long Range Patrol ration first used during the Vietnam conflict, and it is now used by industry for backpacker meals and survival rations, as well as for a variety of food ingredients. Current efforts are focused on cost-effective dehydration processes, such as osmotic dehydration and the development of ready-to-eat, partially dehydrated intermediate moisture items.

The military has continually exploited a variety of mechanical, chemical and electromagnetic energy sources and technologies for purposes of offense, defense or communication.

Numerous electromagnetic energy sources-initially developed for other purposes-can now be used for food and other materials processing. A variety of sources of energy across the electromagnetic spectrum may now be focused on food processing and other materials transformation mation processes. They are a dividend from generations of research aimed at much larger targets than the microorganisms that present threats to stability and safety of food materials. Currently, a considerable variety of thermal and nonthermal processes are being assessed to improve combat ration quality and variety. Among the thermal processes are high-temperature, short-time processes, including ohmic, microwave, radio frequency and induction heating.

The Natick Soldier Center and its precursor, the Quartermaster Food and Container Institute in Chicago, pioneered early research on nonthermal processes. The research explored food preservation potential of ionizing radiation (from both gamma and electron beam sources).

The Natick program was transferred to the USDA in 1980. However, data from the Natick project have been important in getting the approval of low-dose irradiation to improve the safety of fresh meat products and to extend the shelf life of fresh produce. Sterilized irradiated meat items prepared for NASA under Natick Soldier Center guidance are currently used in the Space Shuttle program.

Other pacing technologies envisioned for 21st century combat rations are two nonthermal processes, pulsed electric field (PEF) and high pressure (HP) preservation.

HP preservation can be applied to either liquid or solid foods in flexible containers. The improved food preservation technologies are being used to:

- Minimize processed-induced loss of color, flavour, texture, and nutrition.
- Retain the highest possible quality in stressful storage environments.
- Provide shelf-stable foods with fresh food attributes.
- Optimize ration variety, acceptance and consumption.
- Reduce the logistics burden and cost vs. conventionally-processed foods.
- Enhance the overall quality of life of the service member.

Because nonthermal processing is relatively new compared to conventional processing, two key issues are particularly critical for PEF and HP technologies. The first major issue is safety and process assurance; the second is sensory quality and consumer acceptance over the intended shelf life.

For military use that shelf life must be at least 3 years at 80 degrees Fahrenheit and 6 months at 100 degrees Fahrenheit. Several products made shelf-stable by nonthermal technologies were selected for microbiological and sensory assessment from the output of feasibility study contracts conducted 1992-1995. The contracts are as follows:

- Pulsed Electric Field (PEF)-FoodCo (now PurePulse Technologies), San Diego, Calif. Demonstration products were spaghetti sauce with meat, banana yogurt drink, milk and orange juice, a refrigerated product.
- Hyperbaric Pressure Preservation-Oregon State University and University of Delaware. Demonstration products were Spanish rice, spaghetti with meat sauce, fruit salad and yogurt with peaches.

After the products were shown to be microbiologically safe, sensory screenings were conducted against available, equivalent commercial products, military products or both using standard descriptive and/or consumer acceptance methods. As an example, batch-processed demonstration HP products submitted by Oregon State University were spaghetti with meat sauce and Spanish rice packed in 8-ounce polymeric bowls with aluminum tear-off closures treated in-package at 75,000 psi for 20 minutes. Despite the infancy of HPP processing, consumer panelists appearance, flavour and texture ratings of the spaghetti and rice items met or exceeded benchmark products both initially and after one year of storage at 80 F. A follow-on technology demonstration effort on HP processing was conducted under the leadership of Oregon State University.

One thrust was to optimize throughput and efficiency of a batch high-pressure processing system. Elmhurst Research, Inc., which turned to cannon technology for inspiration, developed a novel design for a batch processing system.

This effort has become a true "guns-to-plowshares" event in that a prototype system is now under development that uses recycled 6-inch cannon tubes as pressure vessels.

The Army's assessment of these novel technologies indicates that they do have considerable technology transfer potential to the food industry in view of the following:

- Consumer demand for minimally-processed foods with maximum nutrient retention.
- Continuing consumer demand for high quality and convenience along with assurance of safety.
- International competition in the industry.

As examples, ohmic (thermal) products are now commercial in Europe and Asia, and HP products are commercial in Japan. As noted earlier, the military is a potential beneficiary of technological advances in food processing to support integrated and focused logistics for battlefield operations and peacekeeping missions.

Finally, the role of the DoD Combat Ration Program is seen as serving as a catalyst for the food processing industry by securing considerable leveraging of resources through academic, industrial, government, and trade, technology or industry association partners in the United States and abroad. In this regard, the DoD has co-sponsored workshops of nonthermal researchers from industry, academia and government to jointly address the key technical barriers to commercialization of PEF and HP processing.

These workshops have led to the formation of a new Division of Nonthermal Processing for the Institute of Food Technologists.

The application of pulsed power technology to food preservation is being facilitated by a Fiscal Year 1999 Dual Use Science and Technology project.

The project for a Pulsed Electric Field Food Processing Consortium will be led by Ohio State University and industrial partners, including PurePulse, Inc.; TetraPak; Kraft; Nestles; and General Mills.

Industrial adoption of advanced food processing technologies will enhance the sustainment of our warfighters and will contribute to a more flexible and responsive logistics system.

Technological Advancements in Dairy Foods

Irradiation, biopreservatives and common sense ensure safety and extend shelflife. Capitol Hill food activists fall into two camps, those protesting advancements and applications in biotechnology and genetic engineering and those accusing processors of permitting unsafe food to enter the marketplace. Food safety is perhaps one of the greatest consumer concerns of the decade. According to a recent Gallup study commissioned by the American Meat Institute (AMI), of the 1,005 randomly selected adults interviewed nationwide, 79% were very or somewhat concerned about the safety of the food they purchased. These consumers are also extremely vulnerable to the activists on Capitol Hill.

Data from the Yankelovich Monitor indicates that from 1988 to 1993 consumer confidence in the media dropped from 53% to 23%, a 56% decrease. While their trust may be decreasing, the media exposure of food safety issues continues. Inevitably, consumers soak up prime time media, and respond accordingly. And as the media continues to report alleged health hazards in the American diet using descriptors such as "toxic," "carcinogenic" and "radioactive," consumers believe that the modern food supply is inherently dangerous due to the presence of synthetic chemicals and use of artificial ingredients and treatments.

The attached chart ranks food hazards, as calculated by FDA. The primary cause of illness from food is naturally occurring pathogens and their by-products. Pesticide residues and food additives are the least dangerous, but unfortunately, the media does not convey this message. For the most part, consumers do not understand that advancements in the areas activists protest are actually advancements toward a safer food supply.

Steps to Safety

"A Study of Consumers' Awareness, Knowledge and Attitudes Toward the Process of Irradiation," completed by AMI, revealed that the more consumers learn about food irradiation, the more they want it. "Spoilage organisms in dairy products are generally radiation-sensitive," explained Edward Remmers, v.p. of the American Council on Science and Health, at the recent Cultured Dairy Products Conference. "Unfortunately, radiation is not acceptable for dairy products promoting live and active cultures—they will be destroyed."

Remmers described the long-term recommendation made by Dr. E. Josephson, former head of U.S. Army's Food Irradiation Project and current adjunct professor of food science at the University of Rhode Island. "Explore irradiation of nonfat/lowfat cultured products at levels of 0.5, 1.0, 1.5 and 3.0 Kilograys to make a 2-3 logarithmic reduction in level of microorganisms in order to achieve a 2-3 fold extension of shelflife," says Remmers. "The greatest area of concern is the effect of radiation on radiation-sensitive sulphur-containing amino acids in milk proteins."

Mansel Griffiths, professor of food science, University of Guelph, Ontario, Canada, says, "Always equate shelflife with safety. And remember, there's no substitute for GMPs of HACCP." He explains that control of storage temperature and water activity are critical for the shelflife of yogurt. "Yogurt stored at 40 [degrees] F has a shelflife of about 35 days. Increase that to 50 [degrees] F, and shelflife reduces to 19 days. And the lower the water activity, the longer the shelflife."

Overcoming High Moisture

Researchers at the University of Wisconsin-Madison Food Research Institute continue to assess the quality and safety of dairy foods. During a recent meeting with Eric Johnson, associate professor, and John Luchansky, assistant professor, both from the department of food microbiology and toxicology; and Mark Johnson, senior scientist at the Center for Dairy Research, they agreed that both starter and nonstarter bacteria flourish in higher moisture, lower fat cheeses.

Growth of undesirable microbes in lower fat cheese results in faster protein breakdown (softer body), gassing and/or off-flavour development. They emphasize concentrating on the following variables in cheese manufacturing, especially in reduced fat versions, to ensure safety and shelflife.

- Pasteurization of milk is an absolute must for reduced-fat and fat-free cheeses.
- Good Manufacturing Practices will keep coliforms out.
- Biofilm development can be prevented by properly cleaning and sanitizing vats, eliminating the transfer and growth of undesirable non-pathogenic bacteria.
- Salt to water ratio must be properly maintained in order to inhibit the growth of nonstarter bacteria. Unfortunately, an increase in salt also inhibits starter bacteria, and slows protein breakdown and body development, resulting in a salty bland cheese. Salt substitutes may prove beneficial.
- Strict control of bacteria in the manufacturing and storage areas can reduce the numbers of non-starter bacteria. Cold storage temperatures can inhibit growth.
- Efficient distribution should be incorporated into the market strategy for higher moisture cheeses due to flavour and texture breakdown.

Natural Preservatives

Luchansky's research group is currently conducting extensive research on the effects of biopreservatives in natural cheese. While this concept of utilizing lactic culture and their metabolites as a preservative has been around for years, it has increased in the last five years due to pressures from activists against other forms of safety assurance.

The bacteriocins (natural preservatives derived from bacteria) evaluated most extensively in dairy foods are nisin (produced by

Lactococcus lactis) and pediocin (produced by Pediococcus acidilactici). "Our group has incorporated bacteriocins at various stages during cheese manufacture to control both pathogenic and spoilage bacteria," says Luchansky. "For example, in the presence of increasing levels of pediocin, numbers of Listeria monocytogenes decreased by 1.5 to 4.0 units compared to control treatments." This experiment involved treated Cheddar cheese samples held at 39.2 [degrees] F for 35 days.

Luchansky suggests selecting or constructing bacteriocin-producing cultures for use as starter or starter adjuncts in the cheese make process. Or, commercially available fermentates of lactic acid bacteria can be added to the milk or to the curd with the salt. Additional research includes bacteriocin applications by means of packaging material and surface application.

Ranking of Food Hazards (Most to Least Dangerous)

1. Microbial contamination
2. Naturally occurring toxicants
3. Environmental contaminants (metals)
4. Nutritional problems (malnutrition, undernutrition)
5. Pesticide residues
6. Food additives.

During the Middle Ages, farmers in North Africa, the Near East, and Europe began making use of agricultural technologies including irrigation systems based on hydraulic and hydrostatic principles, machines such as norias, water-raising machines, dams, and reservoirs. This combined with the invention of a three-field system of crop rotation and the moldboard plow greatly improved agricultural efficiency.

In the European medieval period, agriculture was considered part of the set of *seven mechanical arts.*

Modern Era

After 1492, a global exchange of previously local crops and livestock breeds occurred. Key crops involved in this exchange included the tomato, maize, potato, manioc, cocoa bean and tobacco going from the New World to the Old, and several varieties of wheat, spices, coffee, and sugar cane going from the Old World to the New.

The most important animal exportation from the Old World to the New were those of the horse and dog (dogs were already present in the pre-Columbian Americas but not in the numbers and breeds suited to farm work). Although not usually food animals, the horse (including donkeys and ponies) and dog quickly filled essential production roles on western-hemisphere farms.

The potato became an important staple crop in northern Europe. Since being introduced by Portuguese in the 16th century, maize and manioc have replaced traditional African crops as the continent's most important staple food crops. By the early 19th century, agricultural techniques, implements, seed stocks and cultivated plants selected and given a unique name because of its decorative or useful characteristics had so improved that yield per land unit was many times that seen in the Middle Ages. Although there is a vast and interesting history of crop cultivation before the dawn of the 20th century, there is little question that the work of Charles Darwin and Gregor Mendel created the scientific foundation for plant breeding that led to its explosive impact over the past 150 years.

With the rapid rise of mechanization in the late 19th century and the 20th century, particularly in the form of the tractor, farming tasks could be done with a speed and on a scale previously impossible. These advances have led to efficiencies enabling certain modern farms in the United States, Argentina, Israel, Germany, and a few other nations to output volumes of high-quality produce per land unit at what may be the practical limit.

The Haber-Bosch method for synthesizing ammonium nitrate represented a major breakthrough and allowed crop yields to overcome previous constraints. In the past century agriculture has been characterized by enhanced productivity, the substitution of labour for synthetic fertilizers and pesticides, water pollution, and farm subsidies. In recent years there has been a backlash against the external environmental effects of conventional agriculture, resulting in the organic movement.

The cereals rice, corn, and wheat provide 60% of human food supply. Between 1700 and 1980, "the total area of cultivated land worldwide increased 466%" and yields increased dramatically, particularly because of selectively bred high-yielding varieties, fertilizers, pesticides, irrigation, and machinery. For example, irrigation increased corn yields in eastern Colorado by 400 to 500% from 1940 to 1997. However, concerns have been raised over the sustainability of intensive agriculture. Intensive agriculture has become associated with decreased soil quality in India and Asia, and there has been increased concern over the effects of fertilizers and pesticides on the environment, particularly as population increases and food demand expands. The monocultures typically used in intensive agriculture increase the number of pests, which are controlled through pesticides. Integrated pest management (IPM), which "has been promoted for decades and has had some notable successes" has not significantly affected the use of pesticides because policies encourage the use of pesticides and IPM is knowledge-intensive.

Although the "Green Revolution" significantly increased rice yields in Asia, yield increases have not occurred in the past 15–20 years. The genetic "yield potential" has increased for wheat, but the yield potential for rice has not increased since 1966, and the yield potential for maize has "barely increased in 35 years". It takes a decade or two for herbicide-resistant weeds to emerge, and insects become resistant to insecticides within about a decade. Crop rotation helps to prevent resistances.

Agricultural exploration expeditions, since the late 19th century, have been mounted to find new species and new agricultural practices in different areas of the world. Two early examples of expeditions include Frank N. Meyer's fruit- and nut-collecting trip to China and Japan from 1916-1918 and the Dorsett-Morse Oriental Agricultural Exploration Expedition to China, Japan, and Korea from 1929-1931 to collect soybean germplasm to support the rise in soybean agriculture in the United States.

In 2009, the agricultural output of China was the largest in the world, followed by the European Union, India and the United States, according to the International Monetary Fund. Economists measure the total factor productivity of agriculture and by this measure agriculture in the United States is roughly 2.6 times more productive than it was in 1948.

Six countries - the US, Canada, France, Australia, Argentina and Thailand - supply 90% of grain exports. The United States controls almost half of world grain exports. Water deficits, which are already spurring heavy grain imports in numerous middle-sized countries, including Algeria, Iran, Egypt, and Mexico, may soon do the same in larger countries, such as China or India.

Crop Production Systems

Cropping systems vary among farms depending on the available resources and constraints; geography and climate of the farm; government policy; economic, social and political pressures; and the philosophy and culture of the farmer. Shifting cultivation (or slash and burn) is a system in which forests are burnt, releasing nutrients to support cultivation of annual and then perennial crops for a period of several years. Then the plot is left fallow to regrow forest, and the farmer moves to a new plot, returning after many more years (10-20). This fallow period is shortened if population density grows, requiring the input of nutrients (fertilizer or manure) and some manual pest control. Annual cultivation is the next phase of intensity in which there is no fallow period. This requires even greater nutrient and pest control inputs. Further industrialization lead to the use of monocultures, when one cultivar is planted on a large acreage. Because of the low biodiversity, nutrient use is uniform and pests tend to build up, necessitating the greater use of pesticides and fertilizers. Multiple cropping, in which several crops are grown sequentially in one year, and intercropping, when several crops are grown at the same time are other kinds of annual cropping systems known as polycultures.

In tropical environments, all of these cropping systems are practiced. In subtropical and arid environments, the timing and extent of agriculture may be limited by rainfall, either not allowing multiple annual crops in a year, or requiring irrigation. In all of these environments perennial crops are grown (coffee, chocolate) and systems are practiced such as agroforestry. In temperate environments, where ecosystems were predominantly grassland or prairie, highly productive annual cropping is the dominant farming system.

The last century has seen the intensification, concentration and specialization of agriculture, relying upon new technologies of agricultural chemicals (fertilizers and pesticides), mechanization, and plant breeding (hybrids and GMO's). In the past few decades, a move towards sustainability in agriculture has also developed, integrating ideas of socio-economic justice and conservation of resources and the environment within a farming system. This has led to the development of many responses to the conventional agriculture approach, including organic agriculture, urban agriculture, community supported agriculture, ecological or biological agriculture, integrated farming and holistic management, as well as an increased trend towards agricultural diversification.

Crop Statistics

Important categories of crops include grains and pseudograins, pulses (legumes), forage, and fruits and vegetables. Specific crops are cultivated in distinct growing regions throughout the world. In millions of metric tons, based on FAO estimate.

Livestock Production Systems

Animals, including horses, mules, oxen, camels, llamas, alpacas, and dogs, are often used to help cultivate fields, harvest crops, wrangle other animals, and transport farm products to buyers. Animal husbandry not only refers to the breeding and raising of animals for meat or to harvest animal products (like milk, eggs, or wool) on a continual basis, but also to the breeding and care of species for work and companionship. Livestock production systems can be defined based on feed source, as grassland - based, mixed, and landless. Grassland based livestock production relies upon plant material such as shrubland, rangeland, and pastures for feeding ruminant animals. Outside nutrient inputs may be used, however manure is returned directly to the grassland as a major nutrient source.

This system is particularly important in areas where crop production is not feasible because of climate or soil, representing 30-40 million pastoralists. Mixed production systems use grassland, fodder crops and grain feed crops as feed for ruminant and monogastic (one stomach; mainly chickens and pigs) livestock. Manure is typically recycled in mixed systems as a fertilizer for crops. Approximately 68% of all agricultural land is permanent pastures used in the production of livestock.

Landless systems rely upon feed from outside the farm, representing the de-linking of crop and livestock production found more prevalently in OECD member countries. In the U.S., 70% of the grain grown is fed to animals on feedlots. Synthetic fertilizers are more heavily relied upon for crop production and manure utilization becomes a challenge as well as a source for pollution.

Production Practices

Tillage is the practice of plowing soil to prepare for planting or for nutrient incorporation or for pest control. Tillage varies in intensity from conventional to no-till. It may improve productivity by warming the soil, incorporating fertilizer and controlling weeds, but also renders soil more prone to erosion, triggers the decomposition of organic matter releasing CO_2, and reduces the abundance and diversity of soil organisms.

Pest control includes the management of weeds, insects/mites, and diseases. Chemical (pesticides), biological (biocontrol), mechanical (tillage), and cultural practices are used. Cultural practices include crop rotation, culling, cover crops, intercropping, composting, avoidance, and resistance. Integrated pest management attempts to use all of these methods to keep pest populations below the number which would cause economic loss, and recommends pesticides as a last resort.

Nutrient management includes both the source of nutrient inputs for crop and livestock production, and the method of utilization of manure produced by livestock. Nutrient inputs can be chemical inorganic fertilizers, manure, green manure, compost and mined minerals.

Crop nutrient use may also be managed using cultural techniques such as crop rotation or a fallow period. Manure is used either by holding livestock where the feed crop is growing, such as in managed intensive rotational grazing, or by spreading either dry or liquid formulations of manure on cropland or pastures.

Water management is where rainfall is insufficient or variable, which occurs to some degree in most regions of the world. Some farmers use irrigation to supplement rainfall. In other areas such as the Great Plains in the U.S. and Canada, farmers use a fallow year to conserve soil moisture to use for growing a crop in the following year. Agriculture represents 70% of freshwater use worldwide.

Processing, Distribution, and Marketing

In the United States, food costs attributed to processing, distribution, and marketing have risen while the costs attributed to farming have declined. This is related to the greater efficiency of farming, combined with the increased level of value addition (e.g. more highly processed products) provided by the supply chain. From 1960 to 1980 the farm share was around 40%, but by 1990 it had declined to 30% and by 1998, 22.2%. Market concentration has increased in the sector as well, with the top 20 food manufacturers accounting for half the food-processing value in 1995, over double that produced in 1954. As of 2000 the top six US supermarket groups had 50% of sales compared to 32% in 1992. Although the total effect of the increased market concentration is likely increased efficiency, the changes redistribute economic surplus from producers (farmers) and consumers, and may have negative implications for rural communities.

Crop Alteration and Biotechnology

Crop alteration has been practiced by humankind for thousands of years, since the beginning of civilization. Altering crops through breeding practices changes the genetic make-up of a plant to develop crops with more beneficial characteristics for humans, for example, larger fruits or seeds, drought-tolerance, or resistance to pests. Significant advances in plant breeding ensued after the work of geneticist Gregor Mendel.

His work on dominant and recessive alleles gave plant breeders a better understanding of genetics and brought great insights to the techniques utilized by plant breeders. Crop breeding includes techniques such as plant selection with desirable traits, self-pollination and cross-pollination, and molecular techniques that genetically modify the organism.

Domestication of plants has, over the centuries increased yield, improved disease resistance and drought tolerance, eased harvest and improved the taste and nutritional value of crop plants. Careful selection and breeding have had enormous effects on the characteristics of crop plants.

Plant selection and breeding in the 1920s and 1930s improved pasture (grasses and clover) in New Zealand. Extensive X-ray an ultraviolet induced mutagenesis efforts (i.e. primitive genetic engineering) during the 1950s produced the modern commercial varieties of grains such as wheat, corn (maize) and barley. The Green Revolution popularized the use of conventional hybridization to increase yield many folds by creating "high-yielding varieties".

For example, average yields of corn (maize) in the USA have increased from around 2.5 tons per hectare (t/ha) (40 bushels per acre) in 1900 to about 9.4 t/ha (150 bushels per acre) in 2001. Similarly, worldwide average wheat yields have increased from less than 1 t/ha in 1900 to more than 2.5 t/ha in 1990. South American average wheat yields are around 2 t/ha, African under 1 t/ha, Egypt and Arabia up to 3.5 to 4 t/ha with irrigation.

In contrast, the average wheat yield in countries such as France is over 8 t/ha. Variations in yields are due mainly to variation in climate, genetics, and the level of intensive farming techniques (use of fertilizers, chemical pest control, growth control to avoid lodging).

Genetic Engineering

Genetically Modified Organisms (GMO) are organisms whose genetic material has been altered by genetic engineering techniques generally known as recombinant DNA technology. Genetic engineering has expanded the genes available to breeders to utilize in creating desired germlines for new crops. After mechanical tomato-harvesters were developed in the early 1960s, agricultural scientists genetically modified tomatoes to be more resistant to mechanical handling.

More recently, genetic engineering is being employed in various parts of the world, to create crops with other beneficial traits. New research on woodland strawberry genome was found to be short and easy to manipulate. Researchers now have tools to improve strawberry flavours and aromas of cultivated strawberries as stated in a publication by Nature Genetics.

Herbicide-tolerant GMO Crops

Roundup Ready seed has a herbicide resistant gene implanted into its genome that allows the plants to tolerate exposure to glyphosate. *Roundup* is a trade name for a glyphosate-based product, which is a systemic, nonselective herbicide used to kill weeds. *Roundup Ready* seeds allow the farmer to grow a crop that can be sprayed with glyphosate to control weeds without harming the resistant crop. Herbicide-tolerant crops are used by farmers worldwide.

Today, 92% of soybean acreage in the US is planted with genetically modified herbicide-tolerant plants. With the increasing use of herbicide-tolerant crops, comes an increase in the use of glyphosate-based herbicide sprays.

In some areas glyphosate resistant weeds have developed, causing farmers to switch to other herbicides. Some studies also link widespread glyphosate usage to iron deficiencies in some crops, which is both a crop production and a nutritional quality concern, with potential economic and health implications.

Insect-resistant GMO Crops

Other GMO crops used by growers include insect-resistant crops, which have a gene from the soil bacterium *Bacillus thuringiensis* (Bt), which produces a toxin specific to insects. These crops protect plants from damage by insects; one such crop is Starlink. Another is cotton, which accounts for 63% of US cotton acreage.

Some believe that similar or better pest-resistance traits can be acquired through traditional breeding practices, and resistance to various pests can be gained through hybridization or cross-pollination with wild species. In some cases, wild species are the primary source of resistance traits; some tomato cultivars that have gained resistance to at least nineteen diseases did so through crossing with wild populations of tomatoes.

Costs and Benefits of GMOs

Genetic engineers may someday develop transgenic plants which would allow for irrigation, drainage, conservation, sanitary engineering, and maintaining or increasing yields while requiring fewer fossil fuel derived inputs than conventional crops.

Such developments would be particularly important in areas which are normally arid and rely upon constant irrigation, and on large scale farms. However, genetic engineering of plants has proven to be controversial.

Many issues surrounding food security and environmental impacts have risen regarding GMO practices. For example, GMOs are questioned by some ecologists and economists concerned with GMO practices such as terminator seeds, which is a genetic modification that creates sterile seeds. Terminator seeds are currently under strong international opposition and face continual efforts of global bans.

Another controversial issue is the patent protection given to companies that develop new types of seed using genetic engineering. Since companies have intellectual ownership of their seeds, they have the power to dictate terms and conditions of their patented product. Currently, ten seed companies control over two-thirds of the global seed sales.

Vandana Shiva argues that these companies are guilty of biopiracy by patenting life and exploiting organisms for profit Farmers using patented seed are restricted from saving seed for subsequent plantings, which forces farmers to buy new seed every year. Since seed saving is a traditional practice for many farmers in both developing and developed countries, GMO seeds legally bind farmers to change their seed saving practices to buying new seed every year.

Locally adapted seeds are an essential heritage that has the potential to be lost with current hybridized crops and GMOs. Locally adapted seeds, also called land races or crop eco types, are important because they have adapted over time to the specific microclimates, soils, other environmental conditions, field designs, and ethnic preference indigenous to the exact area of cultivation. Introducing GMOs and hybridized commercial seed to an area brings the risk of cross-pollination with local land races.

Therefore, GMOs pose a threat to the sustainability of land races and the ethnic heritage of cultures. Once seed contains transgenic material, it becomes subject to the conditions of the seed company that owns the patent of the transgenic material.

Modern Agriculture

Modern agriculture is a term used to describe the wide majority of production practices employed by America's farmers. The term depicts the push for innovation, stewardship and advancements continually made by growers to sustainably produce higher-quality products with a reduced environmental impact.

Intensive scientific research and robust investment in modern agriculture during the past 50 years has helped farmers double food production while essentially freezing the footprint of total cultivated farmland.

Safety

The agriculture industry works with government agencies and other organizations to ensure that farmers have access to the technologies required to support modern agriculture practices. Farmers are supported by education and certification programs that ensure they apply agricultural practices with care and only when required.

Sustainability

Technological advancements help provide farmers with tools and resources to help reduce their environmental footprint and to make farming more sustainable.

New technologies have given rise to innovations like conservation tillage, a farming process which helps prevent land loss to erosion, water pollution and enhances carbon sequestration.

The World Bank, the Bill & Melinda Gates Foundation and others have noted that integrated crop management is based on agro-ecological principles and can increase yields while reducing environmental damage.

Affordability

The goal of modern agriculture practices is to help farmers provide an affordable supply of food to meet the demands of a growing population. With modern agriculture, more crops can be grown on less land allowing farmers to provide an increased supply of food at an affordable price.

Food Safety, Labelling and Regulation

Food security issues also coincide with food safety and food labelling concerns. Currently a global treaty, the BioSafety Protocol, regulates the trade of GMOs. The EU currently requires all GMO foods to be labelled, whereas the US does not require transparent labelling of GMO foods. Since there are still questions regarding the safety and risks associated with GMO foods, some believe the public should have the freedom to choose and know what they are eating and require all GMO products to be labelled.

The Food and Agriculture Organization of the United Nations (FAO) leads international efforts to defeat hunger and provides a neutral forum where nations meet as equals to negotiate agreements and debate food policy and the regulation of agriculture. According to Dr. Samuel Jutzi, director of FAO's animal production and health division, lobbying by "powerful" big food corporations has stopped reforms that would improve human health and the environment.

The "real, true issues are not being addressed by the political process because of the influence of lobbyists, of the true powerful entities," he said, speaking at the Compassion in World Farming annual forum.

For example, recent proposals for a voluntary code of conduct for the livestock industry that would have provided incentives for improving standards for health, and environmental regulations, such as the number of animals an area of land can support without long-term damage, were successfully defeated due to large food company pressure.

Environmental Impact

Agriculture imposes external costs upon society through pesticides, nutrient runoff, excessive water usage, and assorted other problems. A 2000 assessment of agriculture in the UK determined total external costs for 1996 of £2,343 million, or £208 per hectare.

A 2005 analysis of these costs in the USA concluded that cropland imposes approximately $5 to 16 billion ($30 to $96 per hectare), while livestock production imposes $714 million. Both studies concluded that more should be done to internalize external costs, and neither included subsidies in their analysis, but noted that subsidies also influence the cost of agriculture to society. Both focused on purely fiscal impacts.

The 2000 review included reported pesticide poisonings but did not include speculative chronic effects of pesticides, and the 2004 review relied on a 1992 estimate of the total impact of pesticides.

Agriculture accounts for 70 per cent of withdrawals of freshwater resources. However, increasing pressure being placed on water resources by industry, cities and the involving biofuels industry means that water scarcity is increasing and agriculture is facing the challenge of producing more food for the world's growing population with fewer water resources.

Scientists are also realising that water resources need to be allocated to maintain natural environmental services, such as protecting towns from flooding, cleaning ecosystems and supporting fish stocks. In the book *Out of Water: From abundance to scarcity and how to solve the world's water problems*, authors Colin Chartres and Samyukta Varma of the International Water Management Institute lay down a six-point plan of action for addressing the global challenge of producing sufficient food for the world with dwindling water resources.

One of the actions they say is required is to ensure all water systems, such as lakes and rivers, have water allocated to environmental flow. A key player who is credited to saving billions of lives because of his revolutionary work in developing new agricultural techniques is Norman Borlaug. His transformative work brought high-yield crop varieties to developing countries and earned him an unofficial title as the father of the Green Revolution.

Livestock Issues

A senior UN official and co-author of a UN report detailing this problem, Henning Steinfeld, said "Livestock are one of the most significant contributors to today's most serious environmental problems". Livestock production occupies 70% of all land used for agriculture, or 30% of the land surface of the planet. It is one of the largest sources of greenhouse gases, responsible for 18% of the world's greenhouse gas emissions as measured in CO_2 equivalents.

By comparison, all transportation emits 13.5% of the CO_2. It produces 65% of human-related nitrous oxide (which has 296 times the global warming potential of CO_2) and 37% of all human-induced methane (which is 23 times as warming as CO_2. It also generates 64% of the ammonia, which contributes to acid rain and acidification of ecosystems.

Livestock expansion is cited as a key factor driving deforestation, in the Amazon basin 70% of previously forested area is now occupied by pastures and the remainder used for feedcrops. Through deforestation and land degradation, livestock is also driving reductions in biodiversity.

Land Transformation and Degradation

Land transformation, the use of land to yield goods and services, is the most substantial way humans alter the Earth's ecosystems, and is considered the driving force in the loss of biodiversity. Estimates of the amount of land transformed by humans vary from 39–50%. Land degradation, the long-term decline in ecosystem function and productivity, is estimated to be occurring on 24% of land worldwide, with cropland overrepresented.

The UN-FAO report cites land management as the driving factor behind degradation and reports that 1.5 billion people rely upon the degrading land. Degradation can be deforestation, desertification, soil erosion, mineral depletion, or chemical degradation (acidification and salinization).

Eutrophication

Eutrophication, excessive nutrients in aquatic ecosystems resulting in algal blooms and anoxia, leads to fish kills, loss of biodiversity, and renders water unfit for drinking and other industrial uses.

Excessive fertilization and manure application to cropland, as well as high livestock stocking densities cause nutrient (mainly nitrogen and phosphorus) runoff and leaching from agricultural land. These nutrients are major nonpoint pollutants contributing to eutrophication of aquatic ecosystems.

Pesticides

Pesticide use has increased since 1950 to 2.5 million tons annually worldwide, yet crop loss from pests has remained relatively constant. The World Health Organization estimated in 1992 that 3 million pesticide poisonings occur annually, causing 220,000 deaths. Pesticides select for pesticide resistance in the pest population, leading to a condition termed the 'pesticide treadmill' in which pest resistance warrants the development of a new pesticide.

An alternative argument is that the way to 'save the environment' and prevent famine is by using pesticides and intensive high yield farming, a view exemplified by a quote heading the Center for Global Food Issues website:

'Growing more per acre leaves more land for nature'. However, critics argue that a trade-off between the environment and a need for food is not inevitable, and that pesticides simply replace good agronomic practices such as crop rotation.

Climate Change

Climate change has the potential to affect agriculture through changes in temperature, rainfall (timing and quantity), CO_2, solar radiation and the interaction of these elements. Agriculture can both mitigate or worsen global warming. Some of the increase in CO_2 in the atmosphere comes from the decomposition of organic matter in the soil, and much of the methane emitted into the atmosphere is caused by the decomposition of organic matter in wet soils such as rice paddies.

Further, wet or anaerobic soils also lose nitrogen through denitrification, releasing the greenhouse gas nitric oxide. Changes in management can reduce the release of these greenhouse gases, and soil can further be used to sequester some of the CO_2 in the atmosphere.

International Economics and Agriculture

Differences in economic development, population density and culture mean that the farmers of the world operate under very different conditions. A US cotton farmer may receive US$230 in government subsidies per acre planted (in 2003), while farmers in Mali and other third-world countries do without.

When prices decline, the heavily subsidized US farmer is not forced to reduce his output, making it difficult for cotton prices to rebound, but his Mali counterpart may go broke in the meantime.

A livestock farmer in South Korea can calculate with a (highly subsidized) sales price of US$1300 for a calf produced. A South American Mercosur country rancher calculates with a calf's sales price of US$120–200 (both 2008 figures). With the former, scarcity and high cost of land is compensated with public subsidies, the latter compensates absence of subsidies with economics of scale and low cost of land.

In the Peoples Republic of China, a rural household's productive asset may be one hectare of farmland. In Brazil, Paraguay and other countries where local legislature allows such purchases, international investors buy thousands of hectares of farmland or raw land at prices of a few hundred US$ per hectare.

Energy and Agriculture

Since the 1940s, agricultural productivity has increased dramatically, due largely to the increased use of energy-intensive mechanization, fertilizers and pesticides.

The vast majority of this energy input comes from fossil fuel sources. Between 1950 and 1984, the Green Revolution transformed agriculture around the globe, with world grain production increasing by 250% as world population doubled. Modern agriculture's heavy reliance on petrochemicals and mechanization has raised concerns that oil shortages could increase costs and reduce agricultural output, causing food shortages.

Modern or industrialized agriculture is dependent on fossil fuels in two fundamental ways: 1) direct consumption on the farm and 2) indirect consumption to manufacture inputs used on the farm. Direct consumption includes the use of lubricants and fuels to operate farm vehicles and machinery; and use of gas, liquid propane, and electricity to power dryers, pumps, lights, heaters, and coolers. American farms directly consumed about 1.2 exajoules (1.1 quadrillion BTU) in 2002, or just over 1 percent of the nation's total energy.

Indirect consumption is mainly oil and natural gas used to manufacture fertilizers and pesticides, which accounted for 0.6 exajoules (0.6 quadrillion BTU) in 2002.

The energy used to manufacture farm machinery is also a form of indirect agricultural energy consumption, but it is not included in USDA estimates of U.S. agricultural energy use. Together, direct and indirect consumption by U.S. farms accounts for about 2 percent of the nation's energy use. Direct and indirect energy consumption by U.S. farms peaked in 1979, and has gradually declined over the past 30 years.

Food systems encompass not just agricultural production, but also off-farm processing, packaging, transporting, marketing, consumption, and disposal of food and food-related items. Agriculture accounts for less than one-fifth of food system energy use in the United States.

Oil shortages could impact this food supply. Some farmers using modern organic-farming methods have reported yields as high as those available from conventional farming without the use of synthetic fertilizers and pesticides. However, the reconditioning of soil to restore nutrients lost during the use of monoculture agriculture techniques made possible by petroleum-based technology takes time.

In 2007, higher incentives for farmers to grow non-food biofuel crops combined with other factors (such as over-development of former farm lands, rising transportation costs, climate change, growing consumer demand in China and India, and population growth) to cause food shortages in Asia, the Middle East, Africa, and Mexico, as well as rising food prices around the globe. As of December 2007, 37 countries faced food crises, and 20 had imposed some sort of food-price controls. Some of these shortages resulted in food riots and even deadly stampedes.

The biggest fossil fuel input to agriculture is the use of natural gas as a hydrogen source for the Haber-Bosch fertilizer-creation process. Natural gas is used because it is the cheapest currently available source of hydrogen. When oil production becomes so scarce that natural gas is used as a partial stopgap replacement, and hydrogen use in transportation increases, natural gas will become much more expensive.

If the Haber Process is unable to be commercialized using renewable energy (such as by electrolysis) or if other sources of hydrogen are not available to replace the Haber Process, in amounts sufficient to supply transportation and agricultural needs, this major source of fertilizer would either become extremely expensive or unavailable. This would either cause food shortages or dramatic rises in food prices.

Mitigation of Effects of Petroleum Shortages

One effect oil shortages could have on agriculture is a full return to organic agriculture. In light of peak-oil concerns, organic methods are more sustainable than contemporary practices because they use no petroleum-based pesticides, herbicides, or fertilizers. Some farmers using modern organic-farming methods have reported yields as high as those available from conventional farming.

Organic farming may however be more labour-intensive and would require a shift of the workforce from urban to rural areas. It has been suggested that rural communities might obtain fuel from the biochar and synfuel process, which uses agricultural *waste* to provide charcoal fertilizer, some fuel *and* food, instead of the normal food vs fuel debate.

As the synfuel would be used on-site, the process would be more efficient and might just provide enough fuel for a new organic-agriculture fusion.

It has been suggested that some transgenic plants may some day be developed which would allow for maintaining or increasing yields while requiring fewer fossil-fuel-derived inputs than conventional crops.

The possibility of success of these programs is questioned by ecologists and economists concerned with unsustainable GMO practices such as terminator seeds, and a January 2008 report shows that GMO practices "fail to deliver environmental, social and economic benefits."

While there has been some research on sustainability using GMO crops, at least one hyped and prominent multi-year attempt by Monsanto Company has been unsuccessful, though during the same period traditional breeding techniques yielded a more sustainable variety of the same crop. Additionally, a survey by the bio-tech industry of subsistence farmers in Africa to discover what GMO research would most benefit sustainable agriculture only identified non-transgenic issues as areas needing to be addressed. Nevertheless, some governments in Africa continue to view investments in new transgenic technologies as an essential component of efforts to improve sustainability.

Chapter 6

Bacteriological Identification and Antibiotic Sensitivity Testing

Clostridia are Gram-Positive rod-shaped spore-forming anaerobic bacteria. Their in vitro growth, identification in the routine diagnostic laboratory and antibiotic sensitivity testing must therefore follow specific rules and use specific materials. In this chapter the general diagnostic methodology, the specific growth requirements, the different orientation and identification assays and the antibiotic sensitivity testing of clostridia are described. In some of the sections, special emphasis will be placed on the growth and identification of Clostridium dificile. 13.1. General methodology Probably the most important prerequisite in the laboratory diagnosis of any clostridial species is a correct clinical sample sent following the strict requirements of anaerobes. Although the clostridial species differ in their requirements for anaerobiosis, with some being even aero-tolerant, all samples should be plated onto appropriate blood agar plates for anaerobic incubation and onto classical blood agar plates and selective agar plates for enterobacteria for comparative aerobic growth. During the incubation period of up to five days the plates should be examined daily for colony growth.

The whole procedure may sometimes take several days and is often not compatible with the field practise, in either human or veterinary medicine. But following the whole procedure is absolutely necessary to prevent the diagnostic laboratory from using identification and antibiotic sensitivity tests for anaerobes on non-strictly anaerobic bacteria. Precise clinical data from the doctor or the veterinarian are also of importance in the identification of the major and minor pathogen clostridia. Clostridia grow best on enriched media, but apart from being anaerobes, the majority of species have no particular growth requirements (with the exception of the intracellular C. piliforme, of course!).

Clostridia grow on commercially available 5% sheep blood agar plates for anaerobes, such as CDC anaerobe blood agar and Schaedler blood agar. Other classical 5% sheep blood agar, such as Columbia, Brucella and Brain-Heart Infusion, must be supplemented with yeast extract, vitamin K1 and haemin to favour growth of anaerobes (Bisping and Amtsberg, 1988; Quinn et al., 1994; Atlas and Park, 2000). Anaerobic storage or pre-incubation of the plates for a few hours in anaerobiosis enhances the chance of isolation of clostridia. Colony swarming can be prevented by growing the bacteria on phenylethyl alcohol blood agar (PEA) or on 4% agar media.

The species of the genus Clostridium can also grow on dozens of selective blood agars for use in medical or food microbiology, several of which are commercially available, with, for instance, aminoglycosides or colistin/nalidixic acid to inhibit coliforms and other Gram negative bacteria, or Columbia with cycloserin for isolation of C. perfringens (Manteca et al., 2001), or the selective media for C. dificile.One of the problems is the variation in the colony morphology and in the zone(s) of haemolysis when the clostridia grow on different media, leading to confusion in the pre-identification criteria. The origin of the blood is also of importance. The zone(s) of haemolysis by bovine isolates of C. perfringens toxin type A are broader and clearer when grown on ox blood agar than on sheep blood agar (Manteca, Mainil and Daube, unpublished data).

Growth is also possible in liquid media. The broth medium must be supplemented with serum and reducing agents, such as thioglycollate or cooked meat broths. An alternative is to add liver pieces (from rabbits) to BHI broth. The broth with a piece of rabbit liver is most probably the best liquid medium (Kaeckenbeeck and Mainil, unpublished), but its preparation is time-and labour-consuming. In most liquid media the clostridia grow best in the reduced portion of the medium, i.e. at the bottom of the tube, close to the piece of meat or of liver. All liquid media must be heated and boiled for 5-10 minutes prior to inoculation, in order to expel the absorbed traces of O_2, followed by cooling at 37°C (Quinn et al., 1994).

Because they are anaerobes, the clostridia were separated early from the genus Bacillus (Mainil et al., 2004). In addition, all clostridia prefer to grow in an atmosphere with 2 to 10% CO_2 (Quinn et al.,

1994). Nevertheless, variations exist in their anaerobic requirements. Some species are very strict anaerobes and do not grow in presence of even traces of O_2. Conversely, other species are aerotolerant (C. histolyticum, C. tertium) and can still grow either in a mixture of 5% O_2, 10% CO_2, 85% N_2, or even in air, though less extensively than in absence of O_2. In contrast with Bacillus species, Clostridium species can sporulate only in absence of O_2 (Peck et al., 2004).

Colony growth normally takes 24h to 48h at a near-neutral pH and 37°C, but may be delayed up to 5 days for some species (C. botulinum, C. colinum) or in nonoptimal conditions. Incubation inside an anaerobic cabinet is the best alternative and is very flexible, allowing daily examination of the plates. But this equipment is of course very expensive. Incubation in anaerobic jars with envelope systems (GasPak) generating anaerobiosis (H_2 + CO_2) is the second best alternative.

However daily examination becomes costly since new packs must be used each time. Moreover, contact with O_2, even shortly, can already reduce the viability of the most strictly anaerobic species. The third alternative is the anaerobic jar with replacement of the air by N_2 and 5-10% CO_2. A catalyst must be present inside the jars to remove the remaining O_2 traces and to allow the growth of strict anaerobes.

However, most often this system does not favour the growth of the most strictly anaerobic clostridia, while favouring the growth of more aerotolerant species, leading to diagnostic confusions and errors.

Liquid media have the advantage of being aerobically incubated. However, one must also be careful, since not all of them support the growth of the most strictly anaerobic species. Whatever system is used, a strictly aerobic bacterial species, like Pseudomonas aeruginosa, and a strictly anaerobic bacterial species should be grown in parallel as controls for anaerobiosis.

Just after the discovery of the pathogenic role of C. dificile, George et al. proposed a selective agar plate called CCFA (cycloserine cefoxitin fructose agar) (George et al., 1979) for isolation from stool specimens of humans, horses and pet animals. This medium containing yeast extract, haemin, vitamin K, and cysteine is still used in most laboratories. The selective agents are cycloserine at the concentration of 500 mg/L and cefoxitine at 16 mg/L. These concentrations have been reduced to 250 mg/L and 10 mg/L respectively in some studies (Levett, 1985). Cefotaxime may also be used instead of cefoxitine at the concentration of 4 mg/L. The original formulation included egg-yolk, which may be replaced by blood. This is commercially available from several companies. Stools are directly inoculated and incubated in an anaerobic atmosphere for 48h. Anaerobic pre-incubation of the plate may also improve the recovery rate (Peterson et al., 1996). The use of an Anoxomat system, which enables the obtention of an anaerobic atmosphere in the jar within one minute, allows the reading of the plates after only a 24 hour incubation period (Delmée, unpublished data). It is worth saying that, in the vast majority of cases, the use of a standard CCFA medium with a standard inoculation procedure is quite satisfactory for the diagnosis of most cases of C. dificile-associated diarrhoea. Culture is the most sensitive method but it is not very specific, due to the possibility of isolating nontoxigenic isolates. It is a slow method, but one that allows the testing of strains for toxigenicity. It is also the only method for epidemiological investigations.

The shapes of the colony and of the haemolysis zone(s) are important orientation criteria. Gram and Ziehl-Neelsen (acid-fast) stained smears must be performed to study the morphology of the cells and of the spores. Any suspect colony must subsequently be subcultured anaerobically, aerobically and microaerobically to prove the requirement for anaerobiosis, before applying the identification tests (without forgetting that a few minor pathogen species are aerotolerant). Identification is performed by study of the biochemical properties, by gas liquid chromatography and by molecular methods, including toxin typing. In strict anaerobic conditions, colonies appear within 24 to 48 hours. Some species are, however, fastidious growers and require longer incubation. The three criteria to examine carefully are the colony morphology (surface, form), the presence/absence of colony swarming and the haemolysis pattern (number and width of haemolysis zones).

The orientation tests are generally similar to those for non anaerobes : Gram and Ziehl-Neelsen stained smears, catalase, cytochrome peroxidase and motility tests. Production of lecithinase and lipase are also considered useful preidentification tests. Though clostridia are Gram-Positive bacteria, Gram stained smears display a mixture of Gram-Positive and Gram-Negative bacterial cells after in vitro growth. In Gram stained smears, most clostridia appear as straight or curved rods, but the shape of a few species can also vary from coccoid rods to long filamentous forms.

The ends of the rods are rounded, tapped or blunt. The cells of many species occur singly or in pairs and several can also form chains of varying lengths and numbers of cells. In chains, the individual cells are of similar length (C. chauvoei) or can be of greatly different lengths (C. septicum). The chains formed by a few species can form tight coils or spiral-shaped (C. spiroforme) configurations. Specific arrangements are, however, more frequent and typical in smears from the lesions than in smears from the colonies. The shape and arrangements of the major pathogen clostridia in Gram stained smears from colonies are of course very important for preidentification, but other minor or non pathogen clostridial species can resemble one or other of the major pathogen species.

In modified Ziehl-Neelsen (acid-fast) stained smears, spores are ovoid or spherical, centrally, subterminally or terminally localized, and cell-distending. Some species sporulate very easily and quickly, while others only rarely. Clostridia give negative results at the catalase (in contrast with Bacillus species) and cytochrome peroxidase tests. Some species can, however, give a weakly positive result at the catalase test. On the other hand, most species are motile, but quickly loose their motility in the presence of O_2.

Most clostridia are saccharolytic or proteolytic, or both, but a few species are neither. The biochemical-based minisystems are best suited for identification of saccharolytic, fast growing clostridia. With these systems the identification can be obtained after 24 to 48hrs of anaerobic incubation by reading the colour reactions manually. A code number is generated and interpreted from the code book or computerized database provided by the manufacturer. The rapid pre-existing enzyme-based minisystems utilise chromogenic or fluorogenic/chromogenic substrates for detection of bacterial glycosidases, aminopetidases and some other enzymes. They are best suited for rapid identification of slow growing clostridia. The inoculation of a heavy bacterial suspension, equal to McFarland standard n° 3 to 5, allows the reading of the results of the reactions after 4 hours of aerobic incubation. The identification can be reached either manually by using a code book or by a computer-assisted automated reader, such as ATB Expression. Nevertheless, none of the identification kits is 100% reliable, even for the identification of the major and minor pathogenic species (60-80%). So, final identification is completed by anti-whole cell antibody immunofluorescent assays (C. chauvoei, C. septicum, C. sordellii, C. novyi, C. haemolyticum), genetic tests, gas liquid chromatography and/or toxin typing assays, if necessary. Colonies of C. dificile are easily recognised on culture plates, due to their typical ground glass appearance when observed under a magnifying glass.A yellow-green or chartreuse fluorescence under ultraviolet illumination is another characteristic of the colonies, but this may vary with the medium used. The typical odour (horse manure) is also an aid to identification. Gas liquid chromatography (GLC) of an agar plot around a suspected colony is the best and easiest method for definite identification. C. dificile displays a typical GLC profile, with large amounts of butyric and iso-caproic acids. Such a device, however, is not available in many laboratories. Individual biochemical tests or anaerobic panels may also be used (rapid pre-existing enzyme-based minisystems are strongly recommended), as well as a somatic antigen latex kit. Cross-reactions, however, have been documented with this latter reagent.

Production of proline-aminopeptidase by a disc test has been recently proposed as a means of rapid identification of C. dificile, when used in conjunction with the typical morphology of the colonies (Delmée, 2001, 2004). A direct enzyme immunoassay (EIA) for toxin A may also be used but then, of course, it will only recognise toxigenic isolates. Enzyme immuno-assays detecting glutamatedehydrogenase (GDH), an enzyme specific to C. dificile, can also help to identify this species, in faeces and after in vitro growth. In combination, these two EIAs for toxin A and for GDH can identify and differentiate non-toxigenic and toxigenic isolates of C. dificile, either after in vitro growth or directly in faeces (TRIAGE C. dificile [Biosite, San Diego, CA]). All or most clostridia species are intrinsically resistant to aminoglycosides (like all anaerobes), sulfonamides (C. perfringens is one exception) and diaminopyrimidines. Acquired resistance to members of other families of antibiotics, especially to the macrolides, lincosamides, tetracyclines and phenicols, has progressively developed at different frequency rates in clostridia species of clinical importance. Several clostridia species can indeed harbour antibiotic resistance genetic determinants on transferable molecules (plasmids, transposons, conjugative transposons) and may serve as reservoir of antibiotic resistance genes for other clostridial species and even genders.

For instance, resistance of human isolates of C. dificile and of human and ruminant isolates of C. perfringens to macrolides and lincosamides is mediated by different erm (Erythromycin Resistance Methylase) genes, possibly originating from Enterococcus or Streptococcus, some of which are located on a conjugative transposon or on a plasmid. Similarly, resistance to tetracyclines is most frequently mediated by the tetA gene amongst human, ruminant and poultry C. perfringens isolates, that can be located on a conjugative plasmid, and by the tetM gene amongst human C. dificile isolates, that can be located on a conjugative transposon which is similar to a conjugative transposon of Enterococcus faecalis. A third example is the resistance of C. perfringens and C. dificile to chloramphenicol, which is mediated by two different CAT (Chloramphenicol Acetyl Transferase) enzymes in C. perfringens, one of which is synthesised by transposon located genes (catP) and is closely related to the catD genes of C. dificile, located on a similar transposon. Different percentages of isolates of other species are also resistant to these three families of antibiotics, but the mechanisms for this are mostly unknown Acquired resistance to other antibiotics, like b lactams (mainly penicillins, more rarely cephalosporins), glycopeptides, fluoroquinolones and metronidazole have also been described. In most species the percentage of resistant isolates remain low. Nevertheless, a high percentage of resistant isolates is found in some minor and major pathogen clostridia. These resistances develop as a consequence of mutations in chromosomal genes or after acquisition of new genes (such as those coding for the production of b lactamase active on penicillins in C. butyricum, C. clostridioforme, C. innocuum and C. ramosum) or are mediated by still unidentified mechanisms.

The determination of the sensitivity/resistance patterns of a clostridial isolate may be important clinically in human medicine, but much less in veterinary medicine. In human hospital medicine, clostridia are tested for antibiotic sensitivity in particular cases, such as the occurrence of several cases of infection or the persistence of an infection in a patient despite instauration of treatment(s) based on the intrinsic sensitivity of the species isolated. Another exception, in veterinary medicine, is probably C. dificile-associated diarrhoea in horses and pet animals. The main problem is the availability of the results only after 3 to 5 days, so that a treatment must be empirically initiated by the clinician. Testing the antibiotic sensitivity patterns of clostridia is therefore important mainly for epidemiological purposes in both human and veterinary medicines, not only for local hospitals, but also because these patterns can vary according to geography following the epidemiology of the specific resistance genetic determinants.

The antibiotics to test are narrow and/or broad spectrum b lactams (penicillins, cephalosporins and carbapenems, with b lactamase inhibitors), glycopeptides (vancomycin, teicoplanin), macrolides (erythromycin), lincosamides (clindamycin), phenicols (chloramphenicol, florfenicol), tetracyclines, recent fluoroquinolones, and/or metronidazole (Prescott et al., 2000). Macrolides can not be tested in anaerobic conditions with CO_2 concentrations between 5 and 10%. It must also be remembered that chloramphenicol (but not florfenicol) was banned years ago for use in farm animals, and that metronidazole has been banned more recently in the European Union. Bacitracin has been used for years in poultry to prevent enterotoxaemia caused by C. perfringens, but is today also banned in the European Union. Nitrofuranes are recommended against C. colinum infections in birds, but are also not allowed in farm animals in the European Union.

A variety of methods is available for susceptibility testing of anaerobic isolates, although problems are encountered with all of them. Agar plate methods include agar dilution, Wadsworth modified agar dilution, spiral gradient endpoint (SGE [Spiral System Instruments, Bethesda, MD]), PDM Epsilometer (E-test [AB Biodisk, Solna, Sweden]) and disk diffusion. Broth methods include broth dilution, broth micro-dilution and broth disk-elution. Not all of these procedures are currently endorsed (e.g. disk diffusion and broth disk-elution) by the NCCLS (National Committee for Clinical Laboratory Standard, 2004) whose guidelines are followed by the majority of the commercially available systems.

Agar dilution, broth macro-and micro-dilution are the techniques suggested to be used in research studies to determine the minimal inhibitory concentrations (MIC) of different antibiotics against a clostridial isolate in anaerobic conditions. The systems used in routine diagnostic laboratories are generally based on broth microdilution methods. There are commercial sources for frozen trays (MSI/Micro-Media System [Cleveland, OH], MicroTech Medical System [Aurora, CO] and Innovative Diagnostic Systems [Atlanta, GA]) or lyophilized trays [Radiometer America/Sensititre, Westlake, OH]). The limited number of appropriate antianaerobic drugs in some of these trays is a disadvantage, but the laboratories can be supplied with custom-poured MIC panels by MicroTech Medical System and by Sensititre.

E-test is another agar-diffusion method used successfully for anaerobes. It is quite easy to perform and to read; it shows a high categorical agreement with the agar dilution reference method in research studies but it is costly. Therefore this method is particularly suited for studying individual isolates against few selected agents.

In human anaerobic bacteriology the MIC values are interpreted and the tested strains categorized (Susceptible, Intermediate, Resistant) according to the breakpoint concentrations recommended either by NCCLS, or by the EUCAST (European Committee on Antimicrobials Susceptibility Testing, 2003), or by national committee guidelines (e.g. Comité de l' Antibiogramme de la Société française de Microbiologie [2005], Deutsches Institute für Normung [2004]) in an attempt to achieve a good predictive value on the basis of in vivo correlations.

Molecular Typing and Identification Methods

The main aim of this chapter is to discuss the molecular characterisation of clostridia. However, some preliminary remarks on molecular identification in general are necessary to provide a background.Classification, characterisation and identification of micro-organisms in general Classification is the theory and process of ordering organisms into recognisable groups. Different classification schemes may exist in parallel. However, today the initial step of classification is mainly based upon characterisation of gene coding sequences for RNA of the small subunit of ribosomes (named 16S rRNA gene in the following pages), the sequences being the basis for unravelling evolutionary lines of descent.

In the pre-molecular era, classification systems were based, for example, on selected cellular and metabolic properties (such as presence or absence of spores and of photosynthetic apparatus) or the relationship to oxygen. Moreover, at the dawn of bacterial systematics, classification was even based on shape, colour or motility.

It is obvious that characterisation and classification are inter-dependent, as the type of classification defines the range of properties to be determined. Present day classification, following the polyphasic approach, is much more demanding than any other previous classification strategy. This is because a broad spectrum of properties needs to be determined at the genetic level (information carried directly on the genes, including non-coding regions) and at the epigenetic level (information carried on the products of the genes).

Bacterial systematics consists of a series of sequential steps. The first step is characterisation, i.e. obtaining data on the properties of an organism. These properties are then compared to those of described species, usually following a dichotomic key or focusing on salient characteristics. Results will show whether the isolate can be identified, i.e., affiliated to a described taxon. If identification is not possible, the isolate is likely to represent a novel taxon which may be described following more extensive characterisation. The extent to which characterisation needs to be performed depends upon the range of properties assigned to phylogenetic neighbouring taxa, to which the new taxon belongs. If sufficient properties have been determined allowing the organisms to be classified (i.e. placed into a systematic framework), a name is given to the appropriate rank of the classified organism. It is named as a new species either of an established genus or of a novel genus. It may be concluded that the purpose and efficiency of a taxonomy is tested by its use in identifying organisms (Stackebrandt et al., 1998).

A chapter on molecular identification methods should not give the impression that non-molecular approaches have lost their significance (even though the molecular approach may one day replace these more traditional methods). In fact, molecular identification constitutes only one, albeit large, segment of a broader spectrum of methods required unambiguously to affiliate isolates to taxa today. Nucleic acid based identification approaches also have obvious advantages over more traditional methods. These are: cost, time, reproducibility, reliable authentification, existence of cumulative databases and their universal application. Nucleic acid based identification approaches are especially advantageous when large strain numbers from the clinical environment need to be identified rapidly, some expressing the most potent toxins known.

Nucleic acid stretches suitable for the identification of members of a given phylogenetic level are molecules showing some of the same characteristics also demanded of suitable phylogenetic markers. The stretches should be present among all representatives of a given group of phylogenetically related organisms in order to enable their study. Since every individual sequence position can only carry information on a rather narrow range of evolutionary time, an increasing number of independently evolving positions or regions raises the number of phylogenetic levels that can be detected.

Depending on the method applied, a nucleic acid stretch should contain at least one region of conserved structure to allow one PCR primer. The conserved region would also have a second primer targeting a taxon-specific variable stretch, to amplify a specific DNA fragment of a wide range of taxa. Between these conserved regions, the nucleotide composition should contain one or more regions of high sequence divergence (e.g., for sequence analyses of intergenic space [ITS] regions and intrageneric stretches). In addition, the rrn operons (coding for rRNA) should contain sequence idiosyncrasies (to allow fluorescent in situ hybridisation [FISH]), together with discriminative sequence analyses and/or electrophoretic separation of DNA fragment of different lengths.

The sequence should contain one, preferably more, restriction sites to allow generation of DNA patterns by Amplified rRNA Digest Restriction Analyses (ARDRA), a 16S rRNA gene-based RFLP method. The availability of the rapidly increasing database of sequenced genes and of completely sequenced genomes will facilitate the comparative search for nucleic acids suitable for identification. Likewise, the following will facilitate the development of identification protocols: the curated database of certain housekeeping genes (MLST, Multi-Locus Sequence Typing), search programs for tandem repeats (MLVA, Multi-Locus Variable-number tandem repeat Analysis) and probe design (ARB software).

Determination of the 16S rRNA gene sequence is among the first steps applied in the identification process. Either purified nucleic acid preparations or crude extracts of bacterial cells can be used for in vitro amplification of rrn genes (= rDNA). In principle, a few bacterial cells are sufficient to perform the analysis. The worldwide application of rDNA gene sequencing makes description of a detailed protocol unnecessary. The reader is referred to the publication by Rainey and Stackebrandt (1992); PCR primer sequences have been indicated by Lane (1991) and updates are regularly included in recent publications on the phylogeny and systematics of prokaryotes. The almost complete gene is amplified using two primers located at the 5' terminus (position 27 forward) and the 3' terminus (position 1525 reverse). A series of internally binding primers (usually 14–20 bases in length) (Lane, 1991) allows rapid sequencing by means of the primer-directed chain termination method.

More than 98% of the Clostridium species type strains have been sequenced (Stackebrandt and Hippe, 2001), including pathogenic species (Stackebrandt et al.,121 1999), the remaining species being unavailable from public collections. Major findings of phylogenetic studies have demonstrated that the genus Clostridium constitutes a broad assembly of distantly related major branches within the Firmicutes (Collins et al., 1994). Studies have also shown that the species C. botulinum, phenotypically circumscribed by the presence of neurotoxins, embraces at least five moderately related strain clusters, each being defined by the presence of a unique neurotoxin. Nearly full length 16S rRNA gene sequences have been subjected to RFLP analysis, digesting the amplificate with restriction enzymes (e.g. TaqI, RsaI), resulting in the discrimination of C. perfringens from other species (Cordoba et al., 2001) and in the classification of pathogenic strains (Gürtler et al., 1991).

As the position of variable regions is known where the majority of speciesspecific nucleotide differences occur, only parts of the complete gene need to be sequenced. Within the 16S rRNA gene, variable regions are located between the following positions: (I) 60 and 106, (II) 146 and 219, (III) 440 and 494, (IV) 588 and 651, (V) 821 and 879, (VI) 997 and 1044, (VII) 1117 and 1154, (VIII) 1240 and 1296, as well as (IX) 1435 and 1466 (E. coli numbering). The degree of variation differs between members of the main lines of descent (phyla and orders).

Amplification of rDNA stretches by polymerase chain reaction (PCR), flanking the variable regions, provides easy access to sequenceable material. The sequences of the members of Clostridium differ mainly in the variable regions located in region I. Long insertions with moderate sequence similarities have been detected in variable region I in Clostridium paradoxum (Rainey et al., 1996) and in members of the genus Desulfotomaculum (Stackebrandt et al., 1997), a neighbouring taxon of Clostridium. The multiple rrn operons may show strong micro-heterogeneity, requiring a cloning step for unambiguous sequence determination.

Regions displaying length heterogeneity are ideal candidates for identification protocols based on electrophoretic analysis of PCR products, especially when species-specific PCR primers are used. Application of two species-specific PCR primers, followed by restriction analysis of one of the resulting fragments, has led to the discrimination of C. estertheticum from other organisms detected in vacuum-packed meat (Helps et al., 1999). Sensitivity can be increased by application of nested PCR, where amplification is achieved by applying two sets of PCR primers, the second one amplifying a subset of the fragments generated by the first set of primers (Klijn et al., 1995)

The identification of highly variable regions with the gene coding for 16S rRNA has led to the development of oligonucleotide probes directed against these mainly species-specific regions. The phylogenetically broader the taxonomic group, the more difficult it is to design a single probe capable of detecting all taxon members. This is especially true for members of the genus Clostridium, which is phylogenetically heterogeneous. A broad group would also detect members of other genera that cluster with certain Clostridium species (e.g. Ruminococcus, Eubacterium, Sarcina), while a species-cluster-specific probe would exclude detection of a major part of the Clostridium species.

Originally, radioactively and digoxigenine labelled probes (Rainey and Stackebrandt, 1992) were designed that hybridised with one of the following: membrane-bound rRNA, isolated DNA or DNA released from lysed filter-fixed colonies. Examples were C. perfringens-specific probes (Rönner et al., 1994a) and probes directed against C. botulinum. Later, fluorescently labelled probes were developed that were able to target RNA directly within the cells. Application of multiple probes carrying different fluorescent labels to the same 16S rRNA, increased sensitivity and allowed detection of non-cultured cells directly within the environment (Amann et al., 1995).

Little data is available for clostridial species, as detection of these organisms in environmental samples has not yet been of high research priority. Examples include the use of fluorescence in situ hybridisation (FISH) probes in the enumeration of human colon bacteria (Harmsen et al., 2000a) and intestinal flora in breast-fed and formula-fed newborn infants (Harmsen et al., 2000b). A combination of a method of PCR amplification of a 16S rRNA gene region, followed by probe hybridisation, increases the sensitivity of detection of Clostridium tyrobutyricum, a bacterium involved in the late blowing of certain cheeses (Klijn et al., 1995).

Some data is available for the 23S rRNA gene, though this molecule is mainly used for phylogenetic studies. Partial 23S rRNA gene sequences have been used to detect C. chauvoei in clinical materials (Sasaki et al., 2000). Using primers directed against the 3'terminus of 16S rDNA and the 5'terminus of 23S rDNA, the intergenic spacer region is accessible for amplification followed by electrophoretic separation of the resulting fragments. One example is the molecular differentiation of some clostridia associated with "blown pack" spoilage (Broda et al., 2003), a species causing the clinical diagnosis in cows of blackleg or malignant oedema. A further example is the identification of intestinal clostridia, including several species pathogenic to humans and animals (Song et al., 2002).

This method has been introduced to base population genetics and epidemiology on a molecular level. Stretches of about 400–500 nucleotides of seven genes, coding for housekeeping proteins, are carefully analysed and differences in alleles recorded. Examples of genes used in MLST are DMSO reductase, aspartokinase glutamin synthase, thymidilate kinase, phosophomannomutase and anthranilate synthase. The selection of genes needs to be evaluated for each new taxon undergoing investigation, as genes common to a particular set of organisms may not be universally distributed. Prominent examples are strain-rich species such as Neisseria gonorrhoea and N. meningitidis, Yersinia pestis and Y. pseudotuberculosis, as well as

Bacillus anthracis and B. cereus. The evolutionary distance between strains is quantified in MLST as the number of differing loci. All strains identical to a particular strain at 5 or more loci are considered members of a clonal complex, or "ecotype". Each species consists of several ecotypes, which should be considered the starting point for independent evolution from within a species as defined today.

The topology of trees of housekeeping genes not involved in recombination is similar to that of trees representing the core of the genome, i.e. genes coding for rRNA. These organisms are considered clonal organisms. In cases where the topologies of these two sets of genes differ, recombination has occurred and these organisms are considered panmictic. Epidemic populations are panmictic, but contain islands of clonal structure, i.e., the population structure is not coherent. No MLST data is as yet available for clostridia, but will very likely be generated because of the availability of strain-rich species with significant pathogenic potential.

The analyses of DNA patterns comprise a broad range of different techniques involving electrophoreses, including restriction fragment analyses and the specific and non-specific generation of PCR-mediated amplification products. This method should be considered a traditional approach. Originally it was used in two-dimensional physical mapping by separation of macro-restriction fragments using pulsed field gel electrophoresis (PFGE). This included a wide spectrum of different variations, for example: CHEF (Contour-clamped homogeneous electric field), FIGE (Field inversion gel electrophoresis), TAFE (Transverse alternating field electrophoresis) and PHOGE (Pulsed homogeneous orthogonal field gel electrophoresis). The advantage of high one-dimensional electrophoretic RFLP resolution was soon evaluated for taxonomic purposes and is used today for the discrimination of highly related strains. The quality of results depends on the generation of high molecular DNA and reproducible separation techniques, as well as on sophisticated software programs to recognise small differences in migration behaviour of fragments. Only a small amount of data is available for members of Clostridium, such as in the study by Cordoba et al. (2001).

This study differentiated C. perfringens from other strains of other Clostridium species. AFLP is the method indexing variations in the whole genome, depending on the presence of restriction sites, and is one which is highly reproducible. The method consists of three steps. Firstly, total DNA is digested using 2 restriction enzymes and the resulting restriction fragments subsequently ligated to halfsitespecific adaptors. Secondly, a subset of the fragments is selectively amplified, using 2 restrictionsite/adaptor-specific primers. Use of fluorescently labelled primers (FAFLP), will refine this method. Thirdly, the PCR products (30–50 fragments of size 80–550 bp) are electrophoretically separated on polyacrylamide gels and fragments numerically evaluated by computer assistant analysis. The AFLP method has been used to detect C. novyi and C. perfringens in drug users (McLauchlin et al., 2002) and to investigate C. perfringens food poisoning in seven outbreaks in the UK. A comparison of PFGE and AFLP has been published by Autio et al. (2003) concerning persistent and sporadic Listeria monocytogenes strains, distant relatives of Clostridium genus members.

The most recent development using DNA patterns for characterisation, identification and epidemiology uses the polymorphism of tandem repeats (TR) as informative markers. Although this technique has not yet been used for differentiation of clostridial strains, it should be briefly mentioned here, as its application to clostridia can be anticipated. Originally found in mammalian cells, TR reminiscent of mini-and micro-satellites have also been found in pathogenic prokaryotes (van Belkum et al., 1997). This is despite the fact that these species constitute genetically very homogeneous entities. The role of TR in the regulation and translation of genes has been discussed by Le Flche et al. (2001), and the position of TRs has been elucidated by analysis of fully sequenced genomes. The data of these authors indicate firstly, that the density of TR array length (< 100 bp) varies widely in prokaryotic genomes (excess of repeats in Buchnera sp., Ureaplasma urealyticum). Secondly, it indicates that some species have a very strong excess of TR, with repeat unit lengths comprising a multiple of three (e.g. Mesorhizobium loti, Mycoplasma tuberculosis and Pseudomonas aeruginosa). A TR database is available at. This locates tandem repeats according to a number of criteria, among others, array length (length of gene fragment enclosed), repeat unit length and number of repeats.

A number of epidemiological molecular subtyping analysis studies have been published, with probably the most well known involving Bacillus anthracis (Keim et al., 2000). Here, two different methods were used to enable typing of strains of B. anthracis, including those involved in bioterrorism. The methods were firstly, sequencing of the protective antigen gene on plasmid pXO1MLVApagA and secondly, Multi-Locus Variable-number tandem repeat Analysis (MLVA). For MLVA analyses, 8 variable number TR markers (6 chromosomal markers and 2 plasmid markers), were combined in 4 multiplex PCR reactions. The PCR products were labelled with fluorescent primers using different coloured dyes, and the amplificates pooled for visualisation on an ABI Genescan. Thus, the location and length of amplificates were easy to detect and to evaluate by applying numerical analyses. The advantage of this system lies in the use of defined DNA regions across the entire genome. The creation of a cumulative database has enabled the use of standardised amplification and separation methods. The disadvantage of this system lies in the need to have sequence information (preferably of more than a single strain of a species) available prior to the design of PCR primers.

With the combination of methods covering the isolation of DNA from any environmental sample (PCR amplification of 16S rRNA genes, cloning and sequence analysis), even as yet uncultured bacteria and microbial communities from natural samples have become accessible for identification and phylogenetic analysis. The presence of cells directly in the environment can be detected by in situ hybridisation techniques (FISH). These studies have demonstrated the enormous taxonomic, hence biotechnological, potential of hitherto unrecognised clostridia in nature.

Gas Chromatography of Metabolic End Products and Cellular Fatty Acids for Identifying Anaerobic Bacteria

Conventional methods of identifying anaerobic bacteria are based on morphology observation, culture and classical biochemical tests. In contrast to aerobes, these techniques are often insufficient to permit definitive identification, and additional investigations are required to complete the diagnosis. One of the most established methods consists of the detection and determination of organic compounds.

Two main groups of compounds may serve as support in the identification of anaerobes:

- Metabolites, produced by bacteria as a result of carbohydrate fermentation and/or breakdown of amino acids.
- Long-chain fatty acids, which are the major constituents of lipids (cell membranes), lipopolysaccharides (Gram-Negative bacteria) and lipoteichoic acid (Gram-Positive bacteria).

Gas chromatography (GC) is a well known technique allowing analysis of these metabolic or structural substances. Thus, GC offers 2 possibilities of additional means for the identification of anaerobes. In this chapter we examine the procedures and potential of each one of these approaches in anaerobic bacteriology. Anaerobic bacteria produce numerous metabolites, which mainly result from sugar fermentation and/or from the breakdown of amino acids by deamination, decarboxylation or Stickland reaction. These metabolites accumulate in the growth medium and are characteristic of bacterial genera and of some species. Most of them have been described in the last 2 decades and consist of 2-oxo-acids, 2-hydroxy-acids, and phenolic compounds (Elsden and Hilton, 1976). Their detection by GC is the most common and most documented technique with regard to the identification of anaerobic bacteria. Nevertheless, some other analytical procedures including HPLC (High-Performance Liquid Chromatography), capillary electrophoresis and indirect UV detection (Arellano et al., 2000) have also been developed for the determination of volatile organic acids. From an analytical point of view, microbial metabolites are classified as volatile fatty acids (VFAs), also called short-chain fatty acids (SCFAs), and as nonvolatile fatty acids (NVFAs). In general, VFAs (acetic to heptanoic), which are the main metabolic products of interest in taxonomy, are directly analysed as free acids after extraction from the growth medium using either diethyl ether or methyl tert-butyl ether. The latter is a safer solvent, since it is considered less flammable.

Other methods have been developed, such as headspace chromatography, a complicated technique used for alcohols (C1–C4) and VFA analysis. Because peak areas are very dependent on analytical conditions (syringe and sample temperature), quantitative analysis is only possible with an automated GC system.

Direct analysis of the supernatants of acidified spent media, faecal or purulent material has also been proposed. A major drawback of this method is that non-volatile compounds tend to decompose in the injector and column. This problem can be solved by pre-treatment procedures or by use of a glass liner in the injector port. This liner is stopped with glass wool plugs and acts as a precolumn. In the last few years, an original extraction technique has been developed, which may be applied to the identification of anaerobic bacteria. Solid-phase microextraction (SPME) is a new solvent-free sample preparation technique, which is inexpensive and time efficient. It utilises a small, fusedsilica fibre coated with a thin layer of polymer to extract analytes from the sample matrix. Then the fibre is introduced into the gas chromatograph and the analytes are thermally desorbed into the injector and subsequently chromatographed. The analysis of fatty acids (C2–C10) with SPME can be carried out directly from liquid (broth culture) or from the headspace above the sample. Although this technique has not yet been applied to bacterial cultures, SPME seems to be a promising method for analysis of VFAs produced by anaerobes. Since this method allows for the making of derivatives, it is certainly also applicable to NVFA analysis. In conclusion, the use of SPME deserves to be developed in anaerobic microbiology.

NVFAs, such as oxo-acids, hydroxylated acids, dicarboxylic acids etc., have a high boiling point, and therefore cannot be directly detected by GC. They must undergo a derivatisation procedure prior to chromatographic analysis. The most commonly used method consists of esterification of analytes by direct incubation of media with methanol-hydrochloric or methanol-sulphuric acid followed by solvent extraction. However, the methylation reaction is incomplete and the recovery rate of methyl esters varies according to the acid. In addition, 2-oxo acids yield at least 2 products 133. A typical example is the methylation of pyruvic acid which produces methylpyruvate as expected, and a second component, identified as methyl-2,2-dimethoxypropanoate. In order to overcome these problems and to obtain reproducible accurate quantitative results, the method has been standardised and a known amount of internal standard (2-methylbenzoic acid) is added in advance to start the methylation reaction (Carlier, 1985).

Esterification of fatty acids can be carried out using different alcohols such as n-butanol, 2-butanol, propanol and methanol. A previous method consisted of converting VFAs and NVFAs to butyl esters with 14 % w/v boron trifluoride – butanol reagent. This method permitted separation of fatty acids (formic acid to heptanoic acid), lactic and succinic acids. Although this technique is no longer used in microbiology, it is always employed for the analysis of fatty acids in dairy products and can be very useful in accurate analysis of VFAs. Indeed, substituting the methyl group with a butyl group improves the flameionisation efficiency of fatty acid esters. Moreover, VFA butyl esters have a uniform flame-ionisation detection (FID) response, irrespective of the number of carbon atoms contained in the molecule (Ulberth et al., 1999). Many other derivatisation procedures can be employed, but their description is beyond the scope of this chapter. (Review: Taguchi, 1990).

The analysis of metabolic end products can be utilised in 2 different ways: firstly, for a rapid presumptive diagnosis of anaerobic infection in the patient, or sometimes for identification by chemical marker of a particular species; secondly, as a tool in the identification of bacteria isolated from a clinical specimen. Proteolytic Clostridium species degrade amino acids (such as valine, leucine, isoleucine, phenylalanine, tyrosine and tryptophan), to iso acids and phenolic compounds. These latter compounds can be considered as microbial markers. Therefore, their detection by headspace analysis or direct injection of body fluids, acidified liquid or solid culture medium, indicates the presence of these proteolytic species in the sample. However, a microscopic examination of the sample is needed, since branched chain acids are also produced by Peptostreptococcus anaerobius and some Gram-Negative bacilli (Porphyromonas, Tissierella). In the same way, a presumptive identification of species can be encouraged by the detection of a specific compound after growth on supplemented broth culture. For example, C. dificile produces p-cresol on cycloserine-cefoxitin-fructose-egg yolk agar (CCFA) or pre-reduced anaerobically sterilised (PRAS) peptone yeast broth supplemented with tyrosine or p-134 hydroxyphenylacetic acid. C. malenominatum, C. tetani, C. tetanomorphum, C. cochlearium and C. limosum produce phenol from tyrosine (Elsden and Hilton, 1976).

Quantitative analysis of metabolic end products in conjunction with some simple tests such as gas, indole or catalase production, allows rapid identification of many common anaerobes. However, care must be taken to obtain reproducible results. Inoculum size, culture medium and time and condition of incubation must be standardised (Carlier, 1985). The more commonly isolated Clostridium may be subdivided into 2 chromatographic groups. The first is composed of species producing one or several straight chain volatile fatty acids, i.e. acetic, or acetic, propionic and butyric acids, whereas the second includes species generating both straight and branched-chain fatty acids.

If acetic acid is the only volatile acid produced, bacteria cannot be identified, since this chromatographic profile is common to many species. Consequently, biochemical tests are required for speciation except for C.135 clostridioforme, which often stains Gram-Negative and forms characteristic elongated cells. Among the species yielding butyric acid, C. perfringens produces propionic and 2-hydroxybutyric acids.

It is easy to recognise by its morphology and the presence of a double zone of haemolysis on a sheep-blood agar plate. Other species belonging to group 1 may be tentatively differentiated by some chromatographic characteristics. C. beijerinckii, unlike C. butyricum, does not produce acetic acid and only synthesises small amounts of lactic acid. C. innocuum exhibits a very specific profile, making its identification very easy without any other test. Two small peaks of acetate and caproate are always detected in addition to prominent peaks of butyric and lactic acids (Carlier, 2003). C. paraputrificum and C. tertium are very similar and must be identified on the basis of biochemical tests. C. tertium is nitrate-positive, whereas C. paraputrificum is not. C. symbiosum, a species morphologically similar to C. clostridioforme, is characterised by the production of butyric acid. Another related species, C. sphenoides, does not produce butyrate, but is indole positive (Holdeman et al., 1977). As seen above, 5 species possess chromatographic profiles similar to that of C. perfringens in their production of propionic and 2-hydroxybutyric acids.

However, they differ in terms of phenol production. Moreover, their cellular morphology is very different. In particular, these bacteria display terminal spores. Group 2 Distinctively, on the basis of aromatic and branched-chain acid production, the bacteria belonging to this group form 3 major subgroups. The first subgroup includes species producing isocaproate and one aromatic acid, namely 3-phenylpropionate, as differential metabolites. Note the higher production of 3-phenylpropionic acid associated with both C. sporogenes and proteolytic C. botulinum species. These species are phenotypically similar. In particular, they are lipase positive and proteolytic (milk-digested). Therefore, precise identification must be performed by a toxicity test. Neither related species, C. bifermentans nor C. sordellii, produces butyric acid, but they are both lecithinase positive and proteolytic. Interspecies differentiation can be carried out using the urease test, which is positive for C. sordellii. However, some urease negative strains have been described (Nakamura et al., 1976), and thus, these particular strains cannot be identified by GC pattern alone.

The second subgroup is formed by species producing mainly isocaproic acid and both aromatic acids, i.e. 2-phenylacetic and 3-phenylpropionic acids. C. mangenoti, is phenotypically very similar to the C. sporogenes/proteolytic C. botulinum group and also yields high levels of 3-phenylpropionic acid. However, it is lipase negative and non-toxigenic. This species is rarely present in clinical specimens. C. dificile is easily identifiable by its characteristic metabolic end products. Indeed, almost all organic acids usually encountered in anaerobes are136 produced by this organism. The particular horse-stable odour, together with nonproteolytic activity complete identification.

The third subgroup comprises only 3 species: C. hastiforme, C. argentinense (formerly C. botulinum type G) and its non-toxigenic counterpart, C. subterminale. It is defined by the presence of isovaleric acid and the absence of 3-phenylpropionic acid. Also, much more importantly, production of 2-phenylacetic distinguishes this subgroup from the others. Production of propionic acid differentiates C. hastiforme from C. argentinense/C. subterminale.

Cellular fatty acid (CFA) composition of micro-organisms has been used extensively for bacterial identification and taxonomic purposes. Variability in number of carbon atoms, number and position of double-bonds, cis or trans isomers and substitution groups are stable parameters, provided that highly standardised analysis and culture conditions are observed.. The usual method consists of lysing bacterial cells by suspension in a saponification mixture, causing the release of their fatty acids, which are converted to methyl ester derivatives (FAME), and then analysed by gas chromatography. Esterification of fatty acids can be carried out by many other procedures involving acid or base-catalysed transesterification. In general, CFA analysis includes saturated, unsaturated, hydroxylated, straight and branched molecules with 9 to 20 carbon atoms. Identification is137 performed by comparison of sample retention times with those of commercially prepared FAME standard mixes. Tentative identification of unknown acids is achieved by measuring their equivalent chain length (ECL) against a homolog of the saturated straight-chain monocarboxylic acids.

The amount of each acid is determined via peak areas as a percentage of the total FAME peak areas. However, this technique is not practical for non-specialised laboratories. A major drawback involves the problem of reproducibility and of comparison of multivariate recordings. Besides, many other compounds other than FAMEs may be detected, making the chromatograms difficult to interpret. For example, many anaerobes contain plasmalogens, which are particular phospholipids possessing a vinyl ether bond at the sn-1 position, instead of the usual ester bond found in fatty acids. In this case, the methylation procedure leads to the formation of aldehydes, which are converted into dimethylacetals (DMAs). These derivatives are practically impossible to separate from most FAMEs. As a consequence, in addition to the need for a highly standardised procedure, the use of an automated apparatus coupled with a database is strongly recommended for effective interpretation of chromatograms and the accurate identification of micro-organisms.

Today, the only commercially available system for this purpose is the Microbial Identification System (MIS) produced by Microbial ID and distributed by Agilent Technologies. The MIS requires bacteria to be streaked on agar plates and incubated for 24 h before the derivatisation procedure and gas chromatography. The software takes into account both the presence or absence of FAMEs and DMAs and the ratio of the acids in the sample.

The chromatographic profiles are compared with those contained in identification libraries by using a similarity index based upon calculation of the distance between the profile of the unknown strain and the mean profile of the closest library entry. A similarity value of 0.5 or greater is considered an acceptable identification, and a value below 0.5 is taken as an uncertain identification. However, the MIS is not very sensitive and 30– 40 mg of cells (wet weight) must be used in the analysis. This is a major drawback for fastidious or slow-growing organisms, requiring a long incubation period. Moreover, because CFA composition of bacteria varies with growth media and culture conditions, strict adherence to the analytical protocol recommended by the manufacturer must be followed to avoid misidentification (Ghanem et al., 1991).

Definition of ECL is as follows: For a specified column and carrier gas, in the range where there is a linear relationship between the molecular weight of the reference saturated straight-chain monocarboxylic methyl esters and logarithm of their retention times, interpolation between the retention times of the reference esters allows determination of an ECL for each compound. This value is independent of experimental conditions (Miwa et al., 1960).

In general, studies demonstrating the utility of CFA analysis by MIS for identification of Clostridium species are lacking. According to Allen et al. (1995), among 216 strains representing 18 species, MIS identified 86% of the strains correctly to species level without supplemental tests (C. perfringens, C. dificile, C. butyricum, C. sporogenes, C. septicum, C. glycolicum and C. symbiosum) and 6% of the strains were identified to species level with the aid of some complementary tests. 3% of the strains, including C. ramosum and C. clostridioforme species, were misidentified to genus level.

Another study showed that 100% of C. dificile strains (n=62) were correctly identified by the system, while misidentification was obtained for one strain of C. butyricum, one strain of C. sordellii and one strain of C. septicum (Anderson et al., 1995). From 686 analyses of 285 C. botulinum strains and related organisms, Ghanem et al. (1991) found 14 CFA groups for toxigenic strains and 6 other CFA groups for non-toxigenic strains. Depending on the toxin type or CFA group, the MIS system permitted a correct identification for 74 to 100% of the analyses, 89 to 100% for the toxigenic strains, and 60 to 100% for the non-toxigenic strains tested. The authors concluded that CFA analysis offers a reasonable alternative to mouse assays for laboratories that do not have facilities for these toxicity tests.

Demonstration of the presence of metabolites in spent media or headspace and analysis of CFA are 2 effective applications of GC for identification of bacteria. Detection of metabolic end products is the most used, simplest and least expensive method. Interpretation of chromatograms can be facilitated by comparison with data contained in numerous publications or by chromatography of reference strains. CFA analysis is an application requiring a highly standardised procedure. Furthermore, it is strongly dependent on the apparatus and software provided by the manufacturer. However, this technique allows the differentiation of more than 140 bacterial compounds (Ghanem et al., 1991), and is considered as an important feature in polyphasic taxonomy (Vandamme et al., 1996).

Chapter 7

Controlling Microbial Quality of Food and Food Standards

Food quality is the quality characteristics of food that is acceptable to consumers. This includes external factors as appearance (size, shape, colour, gloss, and consistency), texture, and flavour; factors such as federal grade standards (e.g. of eggs) and internal (chemical, physical, microbial). Food quality is enforced by the Food Safety Act 1990. Members of the public complain to trading standards professionals, who submit complaint samples and also samples used to routinely monitor the food marketplace to Public Analysts. Public Analysts carry out scientific analysis on the samples to determine whether the quality is of sufficient standard.

Food quality is an important food manufacturing requirement, because food consumers are susceptible to any form of contamination that may occur during the manufacturing process. Many consumers also rely on manufacturing and processing standards, particularly to know what ingredients are present, due to dietary, nutritional requirements (kosher, halal, vegetarian), or medical conditions (e.g., diabetes, or allergics). Besides ingredient quality, there are also sanitation requirements. It is important to ensure that the food processing environment is as clean as possible in order to produce the safest possible food for the consumer. A recent example of poor sanitation recently has been the 2006 North American E. coli outbreak involving spinach, an outbreak that is still under investigation after new information has come to light regarding the involvement of Cambodian nationals.

Food quality also deals with product traceability, e.g. of ingredient and packaging suppliers, should a recall of the product be required.

It also deals with labeling issues to ensure there is correct ingredient and nutritional information.

Recommended Minimum Internal Quality Control in Food Microbiology Testing Laboratories

National Standard Methods, which include standard operating procedures (SOPs), algorithms and guidance notes, promote high quality practices and help to assure the comparability of diagnostic information obtained in different laboratories. This in turn facilitates standardisation of surveillance underpinned by research, development and audit and promotes public health and patient confidence in their healthcare services.

The methods are well referenced and represent a good minimum standard for clinical and public health microbiology. However, in using National Standard Methods, laboratories should take account of local requirements and may need to undertake additional investigations. The methods also provide a reference point for method development. National Standard Methods are developed, reviewed and updated through an open and wide consultation process where the views of all participants are considered and the resulting documents reflect the majority agreement of contributors.

Representatives of several professional organisations, including those whose logos appear on the front cover, are members of the working groups which develop National Standard Methods. Inclusion of an organisation's logo on the front cover implies support for the objectives and process of preparing standard methods. The representatives participate in the development of the National Standard Methods but their views are not necessarily those of the entire organisation of which they are a member.

The performance of standard methods depends on the quality of reagents, equipment, commercial and in-house test procedures. Laboratories should ensure that these have been validated and shown to be fit for purpose. Internal and external quality assurance procedures should also be in place.

Whereas every care has been taken in the preparation of this publication, the Health Protection Agency or any supporting organisation cannot be responsible for the accuracy of any statement or representation made or the consequences arising from the use of or alteration to any information contained in it. These procedures are intended solely as a general resource for practicing professionals in the field, operating in the UK, and specialist advice should be obtained where necessary. If you make any changes to this publication, it must be made clear where changes have been made to the original document. The Health Protection Agency (HPA) should at all times be acknowledged.

The HPA is an independent organisation dedicated to protecting people's health. It brings together the expertise formerly in a number of official organisations.

The HPA aims to be a fully Caldicott compliant organisation. It seeks to take every possible precaution to prevent unauthorised disclosure of patient details and to ensure that patient-related records are kept under secure conditions.

This Guidance Note makes recommendations for the minimum internal quality control (IQC) that food microbiology testing laboratories should implement to meet UKAS requirements.

Accreditation Scheme Requirements

Accreditation schemes stipulate some requirements for IQC. UKAS accreditation requires laboratories to have quality control procedures for monitoring the validity of tests undertaken.

General Requirements

Each member of staff should aim to process at least one IQC sample per year for each UKAS accredited test for which they are trained. This is to demonstrate ongoing competency and to demonstrate that each method is giving satisfactory performance. Staff who are not trained in an entire method should process the parts in which they are competent. Laboratories should also aim to perform IQC tests at a frequency related to their level of sample testing.

Results should be recorded on worksheets. Copies of the worksheets may be stored in the individual members of staff training records or reference made to them to provide traceability.

All unexpected results should be investigated and corrective action taken where required. Investigations and actions should be documented.

Although this Guidance Note recommends preparing suspensions of organisms from reference cultures an alternative to this is to use ready prepared materials deemed suitable for the purpose. Such samples should provide a standardised inoculum at any pre-determined count for the required organisms.

EQA samples can also be used to demonstrate competence and method performance.

Preparation of Serial Dilutions

Serial dilutions are prepared as follows:

- Inoculate 1ml of an overnight broth culture (approximately 109 cfu/ml) of the target organism into 9ml of a suitable diluent (eg maximum recovery diluent) to give a 10-1 dilution. Using a separate pipette for each dilution, prepare serial 10-fold dilutions to 10-7.
- Perform a colony count on the serial dilution to calculate the actual cfu/ml of organisms.

Preparation of Spiked Samples

Spiked samples are prepared as follows:

1. To prepare a spiked sample containing approximately 102 cfu/g add 2ml of the 10-6 d dilution to 225ml of food suspension (225ml of diluent containing 25g of food).
2. To prepare a spiked sample containing approximately 104 cfu/g add 2ml of the 10-4 dilution to 225ml of food suspension (225ml of diluent containing 25g of food).
3. To prepare a spiked sample containing very low numbers (10-100) of cfu/25g add 1ml of the 10-7 dilution to 225ml of food suspension (225ml of diluent containing 25g of food). For detection methods the food is suspended in preenrichment or enrichment media.

Note: any food matrix used to prepare spiked samples should be sterile or have a very low background count or be negative for the target organism.

Note: Preparation of spiked food samples should not be done in the routine food laboratory.

Duplicate Testing

Duplicate spiked or real samples should be set up and tested by different staff if possible. Processing duplicate samples using different staff allows variation between staff to be monitored as well as between different portions of the same sample. One set of results from duplicate samples should be designated in advance to be used for reporting. The other should be used for comparison only. An exception to this may be where duplicate testing forms part of the test protocol agreed with the customer, in which case the mean result should be used. If the results differ then the designated senior member of staff should be notified.

Duplicate counts from enumeration procedures should fall within 0.5 log10.

Replicate Testing

Guidance on replicate testing used to estimate the uncertainty of testing is documented.

Detection Methods

Spiked samples for detection methods are processed as follows:

1. For detection methods such as *Campylobacter*, *Escherichia coli* O157,*Listeria*, *Salmonella* and *Vibrio* species.
   - Prepare a spiked sample containing very low numbers of cfu/g of one of the following organisms:
     - *Campylobacter jejuni* NCTC 11322
     - *Escherichia coli* O157 NCTC 12900
     - *Listeria monocytogenes* NCTC 11994
     - *Salmonella poona* NCTC 4840
     - *Vibrio parahaemolyticus* NCTC 10885.

Note: If a verocytotoxin producing strain of *E.coli* is used then cultures must be handled in containment level 3.

Process according to the specific test method(s) for the target organism. For detection methods, isolation of the target organism indicates that the method has worked satisfactorily. If the organism has not been isolated then the process should be repeated using two or more spiked samples.

Enumeration Methods

Spiked samples for enumeration methods are processed as follows:

1. For enumeration methods such as aerobic colony count, coliforms, *Escherichia coli*, *Listeria* species, and *Staphylococcus aureus*.
   - Prepare a spiked sample containing 102 cfu/g of each of the following organisms:

E. coli NCTC 9001

- *L. monocytogenes* NCTC 11994
- *S. aureus* NCTC 6571.

Process according to the following specified test method(s):

- Aerobic colony count at 30oC/48h
- Enumeration of coliforms/*E. coli*
- Enumeration of Enterobacteriaceae
- Enumeration of *Listeria* species
- Enumeration of *S. aureus*.

For enumeration methods such as *Bacillus cereus* and *Clostridium perfringens*.

Prepare a spiked sample containing 104 cfu/g of each of the following organisms:

- *B. cereus* NCTC 7464
- *C. perfringens* NCTC 8237.

Process according to the following specified test method(s):

- Colony count at 30oC/48h
- Enumeration of *B. cereus* and other *Bacillus* species
- Enumeration of *C. perfringens*.

For enumeration methods such as yeasts and moulds.

Prepare a spiked sample containing 102 cfu/g of each of the following organisms:

- *Saccharomyces cerevisiae* NCPF 3178
- *Aspergillus niger* NCPF 2275.

Process according to the following specified test method:

- Enumeration of yeasts and moulds.

For enumeration methods, counts should be compared with those expected to ensure adequate recovery and check that calculations are correct.

Laboratories should have a range for acceptable counts (e.g. within 0.5 log of expected count) that allows for a decision to be made as to whether the test has passed or failed.

An Overview of the Category Framework for Assessing Raw Milk Products

Food Standards Australia New Zealand (FSANZ) is currently assessing the requirements in the Australia New Zealand Food Standards Code for the sale of raw milk products in Australia through Proposal P1007 Primary Production and Processing Requirements for Raw Milk Products.

As part of Proposal P1007, FSANZ has developed a framework in which to consider the various products that could be considered. This Category Framework Approach defines three categories, taking into account the effect of production methods and the intrinsic characteristic of the final product in assessing the potential food safety risk.

FSANZ has based the development of the Category Framework approach on preliminary results of microbiological risk assessments currently being undertaken by the agency, which have examined production factors and intrinsic properties of selected dairy products.

Processing factors include curd cooking temperature, acidification and storage time. Intrinsic factors include moisture content, acidity and salt concentrations.

Category Framework Approach

The three categories in the framework are defined by the effect that production methods and intrinsic characteristics of the final products have on pathogen survival and growth. If the survival and growth of pathogens is more likely in some products, these products present a greater food safety risk compared to products where pathogen growth and survival is less likely.

While cheese is the major commercial raw milk product being considered, the framework approach will endeavour to achieve outcomes that are applicable to all products, including cultured milk, yoghurt, butter, ice cream etc. Therefore, the scope of this Proposal will examine all activities associated with the production of raw milk products, from on-farm milk production through to retail sale and use by the consumer.

Category 1

Products in this category are defined as those products where intrinsic characteristics and/or processing techniques eliminatepathogens that may have been present in the raw milk. Examples of products in this category would include the extra hard grating cheeses. These cheeses are made from raw milk by heating the curd to greater than 48° C, have low moisture content (<36%) and a long maturation/ripening period – steps that result in the death of pathogens and that lead to cheeses have an equivalent level of safety to pasteurised products.

Category 2

Products in this category are defined as those products where intrinsic characteristics and/or processing techniques may allow the survival of pathogens present in the raw milk, but do not support the growth of these pathogens.

This category would apply to products where there is survival, but not growth, of pathogens that may have been present in the raw milk. In this case, control measures would need to ensure that the raw milk used to produce products is of an appropriate microbiological quality and that processing steps would achieve the necessary critical limits to control pathogens.

To produce these products, potentially a combination of control measures and verification activities will need to be applied to provide an acceptable level of microbial safety for consumption by the general population.

Category 3

Products in this category are defined as those products where intrinsic characteristics and/or processing conditions are likely to allow the survival of pathogens that may have been present in the raw milk and may support the growth of these pathogens.

Category 3 products are those where the processing steps that are applied would not reduce pathogens to an acceptable level. In general, if pathogens are present, they would be expected to multiply during manufacture. This category would include products such as raw drinking milk including raw goat's milk and likely to include high-moisture content cheeses.

Application of Category Framework Approach

Following the completion of the risk assessment and consumer research, FSANZ will use the outcomes to assess whether the nature of the hazards and the level of risk warrant permissions within the categories for the production of raw milk products for sale.

This would involve a variation to current regulatory requirements for the production of milk and dairy products in the Food Standards Code and other industry management interventions.

The appropriate control measures underpinning any permission will be presented in assessments reports prepared during the development of a Primary Production and Processing Standard for Raw Milk Products.

Bacteriological Tests for Milk

Milk testing and quality control is an essential component of any milk processing industry whether small, medium or large scale. Milk being made up of 87% water is prone to adulteration by unscrupulous middlemen and unfaithful farm workers. Moreover, its high nutritive value makes it an ideal medium for the rapid multiplication of bacteria, particularly under unhygienic production and storage at ambient temperatures.

We know that, in order for any processor to make good dairy products, good quality raw materials are essential. A milk processor or handler will only be assured of the quality of raw milk if certain basic quality tests are carried out at various stages of transportation of milk from the producer to the processor and finally to the consumer.

There are a number of standard manuals and text books on milk quality control. However these may not be easily available to the emerging small scale to medium scale processors in Kenya.

For these reasons, the Training Programme for the Small Scale Dairy Sector under project GOK/FAO/TCP/KEN/6611, has assembled this guide on Milk Testing and Quality Control so that it may be used for training and by the private small scale dairy processors.

The methods selected are simple and basic and will suffice the requirements of most milk quality control laboratories of small scale processing units. For the larger plants with bigger laboratories more tests are to be found in the bibliography at the end of this booklet.

Milk Testing and Quality Control

What is Milk Quality Control?

Milk quality control is the use of approved tests to ensure the application of approved practices, standards and regulations concerning the milk and milk products. The tests are designed to ensure that milk products meet accepted standards for chemical composition and purity as well as levels of different micro-organisms.

Why have Milk Quality Control?

Testing milk and milk products for quality and monitoring that milk products, processors and marketing agencies adhere to accepted codes of practices costs money. There must be good reasons why we have to have a quality control system for the dairy industry in Kenya.

The reasons are:

To the Milk Producer : The milk producer expects a fair price in accordance with the quality of milk she/he produces.

The Milk Processor : The milk processor who pays the producer must assure himself/herself that the milk received for processing is of normal composition and is suitable for processing into various dairy products.

The Consumer : The consumer expects to pay a fair price for milk and milk products of acceptable to excellent quality.

The Public and Government Agencies: These have to ensure that the health and nutritional status of the people is protected from consumption of contaminated and sub-standard foodstuffs and that prices paid are fair to the milk producers, the milk processor and the final consumer. All the above-is only possible through institution of a workable quality testing and assurance system conforms to national or internationally acceptable standards.

Chapter 8

Technological Principles of Pasteurization

Food Irradiation

Food irradiation is one of the food processing technologies available to the food industry to control organisms that cause food-borne diseases and to reduce food losses due to spoilage and deterioration. Food irradiation technology offers some advantages over conventional processes. Each application should be evaluated on its own merit as to whether irradiation provides a technical and economical solution that is better than traditional processing methods.

***Table:** Possible causes of spoilage (real or apparent) in canned goods*

Type of food: Acid and high acid foods (canned fruits)

Condition of can	*Action to be taken to identify cause*
Insufficient vacuum or headspace	Check vacuum and headspace in relation to storage temperature and altitude
"Springer" or "flipper"	Cool can to 15°C and check if still domed. Check can for denting, if possible measure headspace volume change brought about by dents, by comparing can volume with volume of a sound can
Hydrogen swell	Check degree of detinning in can especially at the liquid level. Look for scratches or pinholes in lacquer or tin coating. Check if can is still domed on cooling to 15°C.
"Hard" or "soft swell"	Leaker spoilage. Check can for gross seam faults, perforation due to corrosion or damage to seams. Examine contents for signs of spoilage and can interior for detinning at air/product interface.

Applications

For products where irradiation is permitted, commercial applications depend on a number of factors including the demand for the benefits provided, competitiveness with alternative processes and the willingness of consumers to buy irradiated food products. There are a number of applications of food irradiation. For each application it is important to determine the optimum dosage range required to achieve the desired effect. Too high a dosage can produce undesirable changes in texture, colour and taste of foods.

Shelf-life extension. Irradiation can extend the shelf-life of foods in a number of ways. By reducing the number of spoilage organisms (bacteria, mould, fungi), irradiation can lengthen the shelf life of fruits and vegetables. Since ionising radiation interferes with cell division, it can be used as an alternative to chemicals to inhibit sprouting and thereby extend the shelf life of potatoes, onions and garlic. Exposure of fruits and vegetables to ionising radiation slows their rate of ripening. Strawberries, for example, have been found to be suitable for irradiation. Their shelf-life can be extended three-fold, from 5 to 15 days.

Disinfestation. Ionising radiation can also be used as an alternative to chemical fumigants for disinfestation of grains, spices, fruits and vegetables. Many countries prohibit the importation of products suspected of being contaminated with live insects to protect the importing country's agricultural base. With the banning of certain chemical fumigants, irradiation has the potential to facilitate the international shipment of food products.

Global Developments

Consensus on Wholesomeness

In 1980, an FAD/IAEA/WHO Expert Committee reviewed in detail all the accumulated data on food irradiation from the past 40 years.

The Expert Committee concluded that irradiation to an overall dose of 10 kGy (kilograys) presents no toxicological hazard and introduces no special nutritional or microbiological problems, thus establishing the wholesomeness of irradiated foods up to an overall average absorbed dose of 10 kGy.

Data were insufficient to formulate conclusions on applications of food irradiation above 10 kGy. Data on radiation chemistry, nutritional and microbiological aspects of food treated above 10 kGy is currently being compiled.

In 1983, the Codex Alimentarius Commission, an international group that develops global food standards for the FAO and the WHO, incorporated the 1980 Expert Committee's conclusions regarding the wholesomeness of irradiated foods into the Codex General Standard for Irradiated Foods. This proposed international standard was submitted to member countries to accept or to modify according to individual country needs. Currently most countries that allow food irradiation approve its use on a case-by-case basis. The Codex Alimentarius Commission has also adopted a Recommended International Code of Practice for the Operation of Radiation Facilities for the Treatment of Foods. It is intended to serve as a guide for irradiator operators and government regulators.

International Trade

More than 30 countries have given clearances for the use of food irradiation to process some 40 food items and approximately 30 facilities world-wide treat food by irradiation processing. Approvals for additional items are being considered in many countries and many food irradiation facilities are being planned. It was anticipated in 1988 that by 1990 there could be approximately 50 commercial/demonstration irradiators in 25 countries. Table below shows commercial applications of food irradiation to fruits and vegetables by country.

Table: *International commercial applications of radiation for fruit and vegetables*

Country	***Location (application date)***	***Food Commodity***
Argentina	Buenos Aires (1986)	Spinach
Belgium	Fleurus (1981)	Dehydrated vegetables
Brazil	Sao Paulo (1985)	Dehydrated vegetables
Chile	Santiago (1983)	Dehydrated vegetables onions, potatoes
China	Shanghai (1985	Potatoes
Cuba	Havana (1987)	Potatoes, onions
German Dem. Rep	Weideroda (1983)	Onions, garlic
	Spickendorf (1986)	Onions
Japan	Hokkaido (1973)	Potatoes
Korea	Seoul (1985	Garlic powder
Netherlands	Ede (1978)	Dehydrated vegetables
South Africa	Johannesburg (1981)	Dehydrated vegetables
	Tzaneen (1981)	Fruits, onions, potatoes
Thailand	Bangkok (1971)	Onions

Auxiliary Raw Materials

Auxiliary raw materials used in fruit and vegetable processing technologies play a major role in the determination of their physical and chemical characteristics, sensory properties and nutritive value.

Water

Water is one of the essential factors in the activity of the processing centres; according to the final utilisation, water can be classified in three categories:

a. for technological utilisation (when it comes into direct contact with raw materials and enters in the finished product's composition),

b. for steam generators and

c. for receptacle cooling, washing of equipment and general hygiene.

Water for Technological Uses

Water coming into direct contact with the raw materials used for processing (washing, blanching, etc.) or that used as filling liquid of some canned products, must be of drinking water quality in terms of its physico-chemical and microbiological conditions.

More important even than fulfilling drinking water standards, water used for these purposes pests certain specific characteristics related to the technological step or the raw material treated during the processing.

When very hard water is used for blanching vegetables some pecto-calcium and pectomagnesium complexes are formed which starts the hardening of vegetable tissues. This process continues over the pasteurization of the finished product.

When fruit is processed in sugar syrup, the use of hard water for the syrup preparation could induce the formation of a pectin-sugar-acid gel facilitated by the medium pH and presence of calcium salts.

Soft water has negative consequences associated with mineral and hydrosoluble substances and losses during blanching of vegetables.

For some specific products such as peeled tomatoes, green beans and fine texture fruit, the addition of calcium salts (chlorure, sulphate, etc.) is employed to correct for low texture.

The water hardness is an essential factor when used as filling liquid for canned products; ideally, the hardness of the water should be adapted to the raw material species used for canning. Thus a hardness of 3° is good for beans, 5° to 9° for green peas, green beans, and for fruit and vegetables with a tendency to disintegrate should use even harder water.

In the technological process of cucumber and gherkin preservation by natural acidification (lactic fermentation), water hardness has a paramount role. The literature maintains that as far as texture is concerned, the best results are obtained by using very hard water (about 30°); but since high magnesium and salt content has a negative effect on taste, in practice it is recommended that a water hardness of about 10° be used, which gives satisfactory results from both points of view. Oxygen present in water can act as a corrosion factor in metal receptacles but this negative influence can be eliminated by preliminary boiling.

An important factor is pH. Water for canning must be neutral or slightly basic. Acid water plays a major role in corrosion which is evident both on receptacles and on iron or copper equipment, where changes of product colour will be induced. More dangerous is the attack on lead pipes or to the mix used for can sticking and this can render the product toxic. For these reasons an acid water must be neutralised before use.

Water for Steam Generators

Two main conditions must be fulfilled:

- hardness has to be as low as possible, even zero, because precipitation of calcium salts can lead to the formation of encrustations ("crusts") in pipes and on equipment walls. For this reason, water treatment is practiced in the majority of installations;
- from a bacteriological point of view, the iron-bacteria must be eliminated with biological filters or oxidising substances. This is necessary in order to avoid iron hydroxide formed during equipment exploitation to deposit on the inner walls of the pipes.

The elimination of iron-bacteria is also of importance for the water used in processing steps.

Water for Receptacle Cooling and General Hygiene

This should be of drinking water standard. Where this is not available in sufficient quantity, the use of industrial water is acceptable but only for cleaning of production rooms/workshops.

Sweeteners

Sugar

Sugar is the conventional name applied to sucrose. Physically there is icing, granulated and lump sugar. In fruit and vegetable processing, sugar is used only in its granulated form; this quality must be in the form of uniform crystals, white, shining and completely soluble in water.

Concentration of various sugar solutions can be rapidly measured by refractometer reading or with areometers graduated in various ways: Brix, Baumé, etc.

Sugar solubility in water is dependent upon temperature; for example, in order to obtain a saturated solution, one must dissolve 2040 g in one litre of water at 20° C and 4870 g at 100° C. Taking into account this temperature related solubility, in practice the majority of sugar solutions are prepared by heating the water. Water should be as soft as possible because the calcium salts can precipitate on boiling.

Corn Syrup (Liquid Glucose)

Corn syrup is obtained industrially by acid or enzymatic starch hydrolysis, using as starting raw materials maize (corn) or potatoes. In fruit processing, mainly in the production of marmalades, it is possible to use corn syrup. The average composition of this corn syrup is of about 32-40% dextrose (glucose), about 40% dextrins and 18-20% moisture. Sweetening power is 50% compared with sucrose.

In a 10%-20% proportion with sucrose, addition of corn syrup has certain advantages:

a. it improves the shine and texture of marmalade;
b. it prevents "sugaring" defect and
c. it reduces the too sweet taste of finished products obtained with sugar alone.

Salt

Salt is used in order to give to the finished products a specifically salty taste and as a preserving substance. From a chemical point of view the term salt means sodium chloride but in practice the product is never in a pure state. The presence of a significant quantity of magnesium chloride increases the hygroscopicity, gives a bitter taste and can induce corrosion of receptacles.

Food Acids

Acetic acid is in use as solutions of various concentrations which are known under the generic name of vinegar. Vinegar can be obtained:

a) from wine, alcohol, cider, beer, etc. by fermentation;

b) by dilution of acetic acid obtained by dry wood distillation or by synthesis.

From a quality point of view, wine vinegar is preferred, as it has a more pleasant flavour. In order to improve taste, other vinegar types are usually flavoured with spices.

In addition to its spicing and flavouring role, vinegar is used and acts as a preservation agent for some vegetables: cucumbers, acidified vegetables, etc.

- Citric acid
- Tartric acid
- Lactic acid.

Pectic Preparations

In fruit processing there many preparations and mixes known as "pectin" are used as liquid or powder extracts.

From a practical point of view, the pectic preparations are classified as:

- strong pectins obtained from apples or citrus fruit peel; this category gives gels rich in sugar;
- weak pectins which gives gels with low proportion of sugar or even without sugar but with the addition of calcium salts.

The fruit industry uses mainly strong pectins. These preparations are characterised, from a commercial point of view, by the capacity of gelification, expressed in degrees. The degree of gelification represents the quantity of sugar in grams able to be transformed in a standard gel (65% sugar and pH=3) by 1 g pectin.

Intensive Sweeteners

"Calorie-reduced" and "low-calorie foods are widely used and are cornering an increasing share of the market. Sweeteners are making an important contribution to the manufacture of sweet foods in these categories. They make it possible to manufacture sweet products without "sweet" being necessarily synonymous with "high-calorie".

Diabetics need to restrict their intake of sugar and various carbohydrates similar to sugar or avoid them altogether. Sweeteners enable diabetics to enjoy sweet tastes without changing their lifestyle. Sweeteners do not contribute to the development of tooth decay; they do not degrade in the mouth to form the acids which are responsible for caries. Thus sweeteners offer consumers a number of advantageous and favourable properties above and beyond merely reducing calories.

Sunett is the trade mark of Hoechst AG for its high intensity sweetener acesulfame K. As an ingredient, it can be used for sweetening all foods produced industrially or at home, or to produce tabletop sweeteners. Like all other sweeteners, Sunett tastes more intensely sweet than sugar; it is about 200 times sweeter than sugar compared with a dilute stock solution.

Synergism. - Sunett is notable for its pronounced synergism with other sweeteners. The synergistic effect leads as quantitative synergism to an intensification of the overall sweetness and as qualitative synergism to an improved taste. The synergism results in a marked intensification of the sweet taste of the blends, which can amount to 30-50% at usual concentrations.

The favourable properties of Sunett, particularly its synergistic behaviour, can be used advantageously in the Sunett-Multi-Sweetener Concept. For high sweetness levels, blends of Sunett with other sweeteners, e.g. aspartame, are particularly favourable in many applications. The synergistic effect of the blend results in a taste that is particularly pleasant and rounded.

Alone or in blends with other sweeteners, Sunett is used mainly where only sweetness is important and no other properties of the foods are to be affected. This applies particularly to beverages (fruit syrups and juices, carbonated beverages, etc.).

In a number of food products, sweet carbohydrates do not only provide the sweet taste; they have other functions to fulfil; for example, they act as bulking or texturing agents and as preservatives by reducing the water activity. In these types of products, Sunett and other sweeteners cannot be used on their own. They must be combined with other substances which perform the required functions. These may be bulking agents or sugar substitutes (for example polyols: sorbitol, mannitol, xylitol). Sunett can be combined with both groups of substances.

Uses. In carbonated soft drinks blends of Sunett with other sweet-tasting substances are often recommended. Beverages formulated using the Sunett Multi-Sweetener Concept, i.e. blends of Sunett with other sweeteners, are usually preferred because of their particularly well balanced and rounded sweetness profile. The taste of these blends is often superiors to single sweeteners.

Sunett is compatible with sugar and other sweet-tasting carbohydrates, both technically and in terms of taste. It can therefore be used in the production of soft drinks with a reduced sugar content. Drinks based on fructose and Sunett are suitable for diabetics. Soft drinks with a reduced sugar content show an excellent taste quality when Sunett or Sunett blends are added to bring them up to the usual level of sweetness.

Fruit nectars and fruit juice drinks differ from most carbonated drinks in that they contain a noticeable amount of sweet carbohydrates provided by the fruit juice. The amount of carbohydrate may vary depending on the type of fruit and amount of juice used.

Amounts of up to 200 mg/l Sunett are often adequate as a single sweetener for the popular types of fruit nectars. With blends of Sunett and aspartame quantities in the order of 100150 mg/l Sunett and 50 mg/l aspartame or 60-70 mg/l of both Sunett and aspartame are sufficient.

Sunett offers such excellent stability that end products, such as drinks, show no reduction in sweetness performance during normal processing methods and storage periods. Sunett can withstand pasteurization, hot filling and aseptic filling without any loss of sweetness.

Jams and marmalades. Sugar contributes a great deal to the texture and stability of conventional jams and marmalades. For the production of sugar-free products either the sugar must be replaced by comparable amounts of sugar substitutes or some of the functions of the sugar must be taken only by other components, such as suitable gelling agents.

Sugar-free jams and marmalades containing sweeteners are more susceptible to microorganisms than sugar-containing products. The risk of spoilage due to yeast fermentation or moulding can be prevented by pasteurization. However, this is only feasible for small jars which will be quickly consumed once they are opened.

In all other cases, it is advisable to add 0.05-0.1% potassium sorbate as a preservative, wherever this is permitted under the relevant food regulations.

For sugar-free jams and marmalades, concentrations in the range of 500-2000 mg Sunett/kg of the finished product are appropriate. It is advisable to add Sunett in the form of an aqueous stock solution towards the end of the boiling process.

Care must be taken to ensure that Sunett is evenly dispersed throughout the whole batch. Because of the excellent compatibility of Sunett with sugar alcohols, fruit jams and marmalades using these ingredients offer an outstanding taste.

Fruit preserves. Sunett can be used for the production of sugar-free or sugar-reduced fruit preserves. At the normal pH values for fruit preserves Sunett can be added even before pasteurization, as the sweetness is not impaired under the usual thermal treatment conditions. Sunett also withstands the usual storage periods without loss of sweetness.

NutraSweet (r) is the commercial name of aspartame (APM), a new sweetener from G.D. SEARLE & Co. which can be used in the most foods in order to give the same taste as sugar. NutraSweet is about 180 to 200 times sweeter than sucrose (sugar) and this value depends on pH, temperature and the type of flavour.

NutraSweet can be used as mentioned above in a mix with Sunett or alone in all sugar-free or calorie-reduced fruit jams, marmalades and preserves. Like other sweeteners, NutraSweet does not promote tooth decay.

International regulatory status. Both Sunett and NutraSweet are widely accepted by food laws in the majority of countries. The following are the main fruit products where Sunett is an accepted sweetener:

- low-joule prepared jelly: max. 500 mg/kg;
- canned fruit without added sugar: max. 500 mg/kg;
- beverages including calorie-reduced fruit nectars: max. 600 mg/l;
- calorie-reduced jams;
- calorie-reduced and dietetic fruit compotes;
- canned fruit and vegetables, fruit puree, jams and marmalades;
- juices, nectars and juice based beverages;
- jams, marmalades and related products: max. 300 mg/kg.

Packaging Materials

Requirements and Functions of Food Containers

The following are among the more important general requirements and functions of food packaging materials/containers:

a. they must be non-toxic and compatible with the specific foods;
b. sanitary protection;
c. moisture and fat protection;
d. gas and odour protection;
e. light protection;
f. resistance to impact;
g. transparency;
h. tamperproofness;
i. ease of opening;
j. pouring features;
k. reseal features;
l. ease of disposal;
m. size, shape, weight limitations;
n. appearance, printability;
o. low cost;
p. special features.

Primary and Secondary Containers

The terms primary and secondary containers have been used. Some foods are provided with efficient primary containers by nature, such as nuts, oranges, eggs and the like. In packaging these, we generally need only a secondary outer box, wrap, or drum to hold units together and give gross protection.

Other foods such as milk, dried eggs and fruit concentrates often will be filled into primary containers such as plastic liners which are then packaged within protective cartons or drums. In this case the secondary container provided by the carton or drum greatly minimises the requirements that must be met by the primary container.

Except in special instances, secondary containers are not designed to be highly impervious to water vapour and other gases, especially at zones of sealing, dependence for this being placed upon the primary container.Since primary containers by definition are those which come in direct contact with the food, we will be far more concerned with them than with secondary containers.

Hermetic Closure

Two conditions of the greatest significance in packaging are hermetic and non-hermetic closure.

The term hermetic means a container which is absolutely impermeable to gases and vapours throughout its entirety, including its seams. Such a container, as long as it remains intact, will automatically be impervious to bacteria, yeasts, moulds, and dirt from dust and other sources since all of these agents are considerably larger than gas or water vapour molecules.

On the other hand, a container which prevents entry of micro-organisms, in many instances will be non-hermetic. A container that is hermetic not only will protect the product from moisture gain or loss, and from oxygen pick-up from the atmosphere, but is essential for strict vacuum and pressure packaging.

The most common hermetic containers are rigid metal cans and glass bottles, although faulty closures can make them non-hermetic. With very rare exceptions flexible packages are not truly hermetic for one or more of the following reasons.

First, the thin flexible films, even when they do not contain minute pinholes, generally are not completely gas and water-vapour impermeable although the rates of gas and water vapour transfer may be exceptionally slow; second, the seals are generally good but imperfect; and third, even where film materials may be gas- and water-vapour-tight, such as certain gages of aluminium foil, flexing of packages and pouches leads to minute pinholes and crease holes.

Hermetic rigid aluminium containers can be readily formed without side seams or bottom end seams. The only seam then to make hermetic is the top end double seam, which may be closed on regular tin can sealing equipment.

Glass containers are hermetic provided the lids are tight. Lids will have inside rings of plastic or cork. Many glass containers are vacuum packed and the tightness of the cover will be augmented by the differential of atmospheric pressure pushing down the cover.

Crimping of the covers, as in the case of pop bottle caps which operate against positive internal pressure, also can make a gas-tight hermetic seal. But bottles fail more often than cans in becoming non-hermetic.

Protection of Food by Packaging Materials

Films and Foils; Plastics

Films and foils have different values for moisture and gas permeability, strength, elasticity, inflammability and resistance to insect penetration and many of these characteristics depend upon the film's thickness.

Important characteristics of the types of films and foils commonly used in food packaging are given in Table below. For the most part such films are used in the construction of inner containers. Since they are non-rigid, their main functions are to contain the product and protect it from contact with air or water vapour. Their capacity to protect against mechanical damage is limited, particularly when thin films are considered.

These materials can exist in many forms, depending upon such variables as identity and mixture of polymers, degree of polymerisation and molecular weight, spatial polymer orientation, use of plasticisers (softeners) and other chemicals, methods of forming such as casting, extrusion or calendering, etc.

One of the newer classes of plastic materials referred to as copolymers illustrate what can be done with mixtures of the basic units from which plastics are built. The term copolymer refers to a mixture of chemical species in the resin from which films and other forms can be made. The many variations possible make copolymers an important class of plastics to extend the range of useful food packaging applications.

Table: *Properties of packaging films*

Material	***Properties***
Paper	Strength; rigidity; opacity; printability.
Aluminium foil	Negligible permeability to water-vapour, gases and odours; grease proof, opacity and brilliant appearance; dimensional stability; dead folding characteristics.
Cellulose film (coated)	Strength; attractive appearance; low permeability to water vapour (depending on the type of coating used), gases, odours and greases; printability.
Polythene	Durability; heat-sealability; low permeability to water-vapour; good chemical resistance; good low-temperature performance.
Rubber hydrochloride	Heat-sealability; low permeability to water vapour, gases, odours and greases; chemical resistance.
Cellulose acetate	Strength; rigidity; glossy appearance; printability; dimensional stability.

Contd...

Material	***Properties***
Vinylidene chloride copolymer	Low permeability to water vapour, gases, odours and greases; chemical resistance; heat-sealability.
Polyvinyl chloride	Resistance to chemicals, oils and greases; heat-sealability.
Polyethylene terephtalate	Strength; durability; dimensional stability; low permeability to gases, odours and greases.

Source: FAD/WFP, (1970)

Plastic Sheets

- Cellophane paper can be used for packing of dried products, mainly for dried fruit leathers.
- Polyethylene sheets have a variety of uses. They are flexible, transparent and have a perfect resistance to low temperatures and impermeability to water vapour. An important advantage is that these sheets can be easily heat-sealed. Utilisation is in forms of sheets and bags.

  It is a good packing material for primary protection of dehydrated products. If a good protection is needed to prevent flavour and gas losses, it will be necessary to combine polyethylene with other materials.

Receptacles and Packagings in Plastic Materials

In this class there are three categories:

a. receptacles that can be heat treated: boxes, bottles and bags. Sterilisable bags used up to 120° C can be manufactured from same raw materials as described under plastic sheets and up to 100° C from cellophane. Polyethylene bags could be used to some extent for packing and pasteurization of sauerkraut.

b. receptacles that are not heat treated during processing of fruit and vegetables, also divided in bags and boxes. Bags are the most used type of packing from plastic materials and they are manufactured from polyethylene or cellophane; an important utilisation is for dried/dehydrated fruits and vegetables.

c. special packagings - which can be contacted (Criovac type) by action of heat once the finished product is already inside the pack and the air is evacuated.

Laminates

Various flexible materials such as papers, plastic films, and thin metal foils have different properties with respect to water vapour transmission, oxygen permeability, light transmission, burst strength, pin holes and crease hole sensitivity, etc. and so multi-layers or laminates of these materials which combine the best features of each are used.

Commercial laminates containing up to as many as eight different layers are commonly custom designed for a particular product.

Laminations of different materials may be formed by various processes including bonding with a wet adhesive, dry bonding of layers with a thermoplastic adhesive, hot melt laminating where one or both layers exhibit thermoplastic properties, and special extrusion techniques. Such structured plastic films may be complete in themselves or be further bonded to papers or metal foils to produce more complex laminates.

Glass Containers

As far as food packaging is concerned, glass is chemically inert, although the usual problems of corrosion and reactivity of metal closures will of course apply. The principal limitation of glass is its susceptibility to breakage, which may be from internal pressure, impact, or thermal shock, all of which can be greatly minimised by proper matching of the container to its intended use and intelligent handling practices. Here consultation with the manufacturer cannot be over-stressed.

The heavier a jar or bottle for a given volume capacity the less likely it is to break from internal pressure. The heavier jar, however, is more susceptible to both thermal shock and impact breakage. Greater thermal shock breakage of the heavier jar is due to wider temperature differences which cause uneven stress between the outer and inner surfaces of the thicker glass. Greater impact breakage susceptibility of the heavier jar is due to the lower resilience of its thicker wall.

Coatings of various types can markedly reduce each of these types of breakage. These coatings, commonly of special waxes and silicones, lubricate the outside of the glass. As a result, impact breakage is lessened by bottles and jars glancing off one another rather than sustaining direct hits when they are in contact in high speed filling lines.

Surface coating after annealing protects glass surfaces from many of the minute scratches appearing in normal handling after annealing ovens; surface coating also improves the high gloss appearance of glass containers and is said to decrease the noise from glass to glass contact at filling lines.

With regard to thermal shock, it is good practice to minimise temperature differences between the inside and outside of glass containers whenever possible. Some manufacturers will recommend that a temperature difference of 44° C (80° F) between the inside and the outside not be exceeded. This requires slow warming of bottles before use for a hot fill and partial cooling before such containers are placed under refrigeration.

Classification

Glass used for receptacles in fruit and vegetable processing is a carefully controlled mixture of sand, soda ash, limestone and other materials made molten by heating to about 1500° C (2800° F).

Main classes of glass receptacles are:

a. jars which are resistant to heat treatments,
b. jars, glasses, etc. for products not submitted to heat treatment (marmalades, acidified vegetables, etc.);
c. glass bottles for pasteurized products (tomato juice, fruit juices, etc.) or not pasteurized (syrups) and
d. receptacles with higher capacity (flasks, etc.)

Jars for Sterilised/Pasteurized Canned Products

These receptacles may replace metal cans, taking into considering both the advantages and disadvantages they present. Advantages are: they do not react to food content; they are transparent and can be manufactured in various shapes; they use cheap raw materials and are reusable. Disadvantages: heavier than metal can of same capacity; fragile; lower thermal conductance and a limited resistance to thermal shocks.

Receptacles in this category must assure a perfect hermeticity after their pasteurization/sterilisation and cooling and this has to be achieved by the use of metallic (or glass) caps and specific materials for tightness. Taking into account the receptacles' closure method, there are two categories of receptacles:

a. glass jars with mechanical closure;
b. glass jars with pneumatic closure.

Jars for Products without Heat Treatment

For marmalades, jellies and jams glass jars with non hermetic closures using metal, glass or rigid plastic caps are used; however for these products the receptacles mentioned above may also be used.

The use of jars with pneumatic closure presents the advantage that some products (e.g. marmalades, jams) can be filled hot and therefore sterile in receptacles. Pneumatic closing generally protects against negative air action which is in this case eliminated from receptacles.

Glass Bottles

These receptacles are widely used both for

a) finished products which need pasteurization (ea. tomato juice, Knit juices, etc.) and for

b) those which are preserved as such (ea. fruit syrups).

Glass bottles in category a) are closed hermetically with metallic caps, provided with special materials for tightness. For glass bottles in category b) various corks, and aluminium caps with tightness materials may be used.

Glass Receptacles with High Capacity

In this category there are glass flasks with 3 and 10 litre capacity which can be hermetically closed by a SKO caps system and are resistant to product pasteurization (ea. tomato juice). As bigger receptacles it is possible to use glass demijohns with usual capacity of 25 and 50 litre; these receptacles are used for preservation of fruit juices by warm process. Closing is performed with flexible rubber hoods.

Paper Packaging

As primary containers few paper products are not treated, coated or laminated to improve their protective properties. Paper from wood pulp and reprocessed waste paper will be bleached and coated or impregnated with such materials as waxes, resins, lacquers, plastics, and laminations of aluminium to improve water vapour and gas impermeability, flexibility, tear resistance, burst strength, wet strength, grease resistance, sealability, appearance, printability, etc.

Paper Sheets

- Kraft paper is the brown unbleached heavy duty paper commonly used for bags and for wrapping; it is seldom used as a primary container;

- parchment paper: acid treatment of paper pulp modifies the cellulose and gives water and oil resistance and considerable wet strength to this type of packaging material;
- glassine-type papers are characterised by long wood pulp fibres which impart increased physical strength;
- paper with plastic material sheets.

Receptacles from Paper or Cardboard

(paper = 8 to 150 g/m^2; cardboard = 150 to 450 g/m^2).

"Tin can"/tinplate

The "tin can" is a container made of tinplate.

Tinplate, a rigid and impervious material, consists of a thin sheet of low carbon steel coated on both sides with a very thin layer of tin. It can be produced by dipping sheets of mild steel in molten tin (hot-dipped tinplate) or by the electro-deposition of tin on the steel sheet (electrolytic tinplate). With the latter process it is possible to produce tinplate with a heavier coating of tin on one surface than the other (differentially coated).

Tin is not completely resistant to corrosion but its rate of reaction with many food materials is considerably slower than that of steel. The effectiveness of a tin coating depends on:

a. its thickness which may vary from about 0.5 to 2.0 µm (20 to 80 x 10(-6) in.);
b. the uniformity of this thickness;
c. the method of applying the tin which today primarily involves electrolytic plating;
d. the composition of the underlying steel base plate;
e. the type of food, and
f. other factors.

Some canned vegetables including tomato products actually owe their characteristics flavours to a small amount of dissolved tin, without which these products would have an unfamiliar taste. On the other hand, where tin reacts unfavourably with a particular food the tin itself may be lacquer coated.

The thickness of tinplate sheets may vary from 0.14 mm to 0.49 mm and is determined by weighing a sheet of known area and calculating the average thickness.

Tinplate sheets may be lacquered after fabrication to provide an internal or external coating to protect the metal surface from corrosion by the atmosphere or through reaction with the can contents. They may also be printed by lithography to provide suitable instructions or information on containers fabricated from tinplate sheets (otherwise paper labels can be attached to the outer tinplate surface).

Under normal conditions the presence of the tin coating provides a considerable degree of electrochemical protection against corrosion, despite the fact that in both types of tinplate the tin coating is discontinuous and minute areas of steel base plate are exposed. With prolonged exposure to humid conditions, however, corrosion may become a serious problem.

The coatings not only protect the metal from corrosion by food constituents but also protect the foods from metal contamination, which can produce a host colour and flavour reactions depending upon the specific food. Particularly common are dark coloured sulphides of iron and tin produced in low acid foods that liberate sulphur compounds when heat processed, and bleaching of red plant pigments in contact with unprotected steel, tin, and aluminium.

Table: *General types of can coatings*

Coating	*Typical uses*	*Type*
Fruit enamel	Dark coloured berries, cherries and other fruits requiring protection from metallic salts	Oleoresinous
C-enamel	Corn, peas and other sulphur-bearing products	Oleoresinous w. suspended zinc oxide
Citrus enamel	Citrus products and concentrates	Modified oleoresinous
Beverage can enamel	Vegetable juices; red fruit juices; highly corrosive fruits; non-carbonated beverages	Two-coated w. resinous base coat and vinyl top coat

Fruit Specific Preservation Technologies

Fruit Quality

Fruit quality goes back to tree stock, growing practices and weather conditions. Closer to the shipper and processor, however, are the degrees of maturity and ripeness when picked and the method of picking or harvesting.

There is a distinction between maturity and ripeness of a fruit. Maturity is the condition when the fruit is ready to eat or if picked will become ready to eat after further ripening. Ripeness is that optimum condition when colour, flavour and texture have developed to their peak.

Some fruit is picked when it are mature but not yet ripe. This is especially true of very soft fruit like cherries and peaches, which when fully ripe are so soft as to be damaged by the act of picking itself. Further, since many types of fruit continue to ripen off the tree, unless they were to be processed quickly, some would become overripe before they could be utilised if picked at peak ripeness.

From a technological point of view, fruit characterisation by species and varieties is performed on the basis of physical as well chemical properties: shape, size, texture, flavour, colour/pigmentation, dry matter content (soluble solids content), pectic substances, acidity, vitamins, etc.

These properties are directly correlated with fruit utilisation. Table below shows which of the above mentioned properties have a major impact on the finished products obtained by fruit processing.

Table: *Optimal use of fruits as a function of their properties*

Processed	***Organoleptic (Sensory) Properties***				***Chemical Composition***		
Finished Products	***Shape***	***Texture***	***Flavour***	***Taste***	***Acidity***	***Sugars***	***Pectic Sust.***
Dried Fruits	+ +	+ +		+ +		+ +	
Fruit Juices			+ +	+ +	+ +		
Marmalade			+ +	+ +			+ +
Jams	+ +	+ +	+ +	+ +			
Jellies	+ +	+ +	+ +	+ +			+ +
Fruit Paste				+ +		+ +	

When to Pick

The proper time to pick fruit depends upon several factors; these include variety, location, weather, ease of removal from the tree (which change with time), and purpose to which the fruit will be put.

For example, oranges change with respect to both sugar and acid as they ripen on the tree; sugar increases and acid decreases. The ratio of sugar to acid determines the taste and acceptability of the fruit and the juice.

For this reasons, in some countries there are laws that prohibit picking until a certain sugar-acid ratio has been reached.

In the case of much fruit to be canned, on the other hand, fruit is picked before it is fully ripe for eating since canning will further soften the fruit.

Quality Measurements

Many quality measurements can be made before a fruit crop is picked in order to determine if proper maturity or degree of ripeness has developed.

Colour may be measured with instruments or by comparing the colour of fruit on the tree with standard picture charts.

Texture may be measured by compression by hand or by simple type of plungers.

As fruit mature on the tree its concentration of juice solids, which are mostly sugars, changes. The concentration of soluble solids in the juice can be estimated with a refractometer or a hydrometer. The refractometer measures the ability of a solution to bend or refract a light beam which is proportional to the solution's concentration. A hydrometer is a weighted spindle with a graduated neck which floats in the juice at a height related to the juice density.

The acid content of fruit changes with maturity and affects flavour. Acid concentration can be measured by a simple chemical titration on the fruit juice. But for many fruits the tartness and flavour are really affected by the ratio of sugar to acid.

Percentage of soluble solids, which are largely sugars, is generally expressed in degrees Brix, which relates specific gravity of a solution to an equivalent concentration of pure sucrose.

In describing the taste of tartness of several fruits and fruit juices, the term "sugar to acid ratio" or "Brix to acid ratio" are commonly used. The higher the Brix the greater the sugar concentration in the juice; the higher the "Brix to acid ratio" the sweeter and lees tart is the juice.

Harvesting and Preprocessing

The above and other measurements, plus experience, indicate when fruit is ready for harvesting and subsequent processing.

A large amount of the harvesting of most fruit crops is still done by hand; this labour may represent about half of the cost of growing the fruit. Therefore, mechanical harvesting is currently one of the most active fields of research for the agricultural engineer, but also requires geneticists to breed fruit of nearly equal size, that matures uniformly and that is resistant to mechanical damage.

A correct manual harvesting includes some simple but essential rules:

- the fruit should be picked by hand and placed carefully in the harvesting basket; all future handling has to be performed carefully in order to avoid any mechanical damage;
- the harvesting basket and the hands of the harvester should be clean;
- the fruit should be picked when it is ready to be able to be processed into a quality product depending on the treatment which it will undergo.

It is worth emphasising the fact that the proximity of the processing centre to the source of supply for fresh raw materials presents major advantages; some are as follows:

- possibility to pick at the best suitable moment;
- reduction of losses by handling/transportation;
- minimises raw material transport costs;
- possibility to use simpler/cheaper receptacles for raw material transport.

Once it has left the tree, the organoleptic properties, nutritional value, safety and aesthetic appeal of the fruit deteriorates in varying degrees.

The major causes of deterioration include the following:

a. growth and activity of micro-organisms;
b. activities of the natural food enzymes;
c. insects, parasites and rodents;
d. temperature, both heat and cold;
e. moisture and dryness;
f. air and in particular oxygen;
g. light and
h. time.

The rapidity with which foods spoil if proper measures are not taken is indicated in Table below.

Table: Useful storage life of some food products

Food Products	*Generalized Storage Life (Days) at 21°C (70°F)*
Animal Flesh, Fish, Poultry	1-2
Dried, salted, smoked meat and fish	360 and more
Fruits	1-7
Dried Fruits	360 and more
Leafy Vegetables	1-2
Root Crops	7-20

Source: Desrosier and Desrosier (1977)

Reception - Quality and Quantity

Fruit reception at the processing centre is performed mainly for following purposes:

- checking of sanitary and freshness status;
- control of varieties and fruit wholeness;
- evaluation maturity degree;
- collection of data about quantities received in connection to the source of supply: outside growers/farmers, own farm.

Variety control is needed in order to identify that the fruit belongs to an accepted variety as not all are suitable for different technological processes.

Fruit maturity degree is significant as industrial maturity is required for some processing/preservation methods while for others there is the need for an edible maturity when the fruit has full taste and flavour.

Special attention is given to size, appearance and uniformity of fruit to be processed, mainly in the form of fruit preserved with sugar using whole/half fruits ("with fruit pieces").

Some laboratory control is also needed, even if it not easy to precisely establish the technological qualities of fruit because of the absence of enough reliable rapid analytical methods able to show eventual deterioration.

The only reliable method for evaluating the quality is the combination of data obtained through organoleptic/taste controls and by simple analytical checks which are possible to perform in a small laboratory: percentage of soluble solids by refractometer, consistency/ texture measured with simple penetrometers, etc.

Temporary Storage before Processing

This step has to be as short as possible in order to avoid flavour losses, texture modification, weight losses and other deterioration that can take place over this period.

Some basic rules for this step are as follows:

- keep products in the shade, without any possible direct contact with sunlight;
- avoid dust as much as possible;
- avoid excessive heat;
- avoid any possible contamination;
- store in a place protected from possible attack by rodents, insects, etc.

Cold storage is always highly preferred to ambient temperature. For this reason a very good manufacturing practice is to use a cool room for each processing centre; this is very useful for small and medium processing units as well.

Washing

Harvested fruit is washed to remove soil, micro-organisms and pesticide residues.

Fruit washing is a mandatory processing step; it would be wise to eliminate spoiled fruit before washing in order to avoid the pollution of washing tools and/or equipment and the contamination of fruit during washing.

Washing efficiency can me gauged by the total number of micro-organisms present on fruit surface before and after washing - best result are when there is a six fold reduction. The water from the final wash should be free from moulds and yeast; a small quantity of bacteria is acceptable.

Fruit washing can be carried out by immersion, by spray/showers or by combination of these two processes which is generally the best solution: pre-washing and washing.

Some usual practices in fruit washing are:

- addition of detergents or 1.5% HCl solution in washing water to remove traces of insect-fungicides;
- use of warm water (about 50°C) in the pre-washing phase;
- higher water pressure in spray/shower washers.

Washing must be done before the fruit is cut in order to avoid losing high nutritive value soluble substances (vitamins, minerals, sugars, etc.).

Sorting

Fruit sorting covers two main separate processing operations:

a. removal of damaged fruit and any foreign bodies (which might have been left behind after washing);

b. qualitative sorting based on organoleptic criteria and maturity stage.

Mechanical sorting for size is usually not done at the preliminary stage. The most important initial sorting is for variety and maturity. However, for some fruit and in special processing technologies it is advisable to proceed to a manual dimensional sorting (grading).

Trimming and Peeling (Skin Removal)

This processing step aims at removing the parts of the fruit which are either not edible or difficult to digest especially the skin. Up to now the industrial peeling of fruit and vegetables was performed by three procedures:

a. mechanically;

b. by using water steam;

c. chemically; this method consists in treating fruit and vegetables by dipping them in a caustic soda solution at a temperature of 90 to 100° C; the concentration of this solution as well as the dipping or immersion time varying according to each specific case.

Cutting

This step is performed according to the specific requirements of the fruit processing technology.

Heat Blanching

Fruit is not usually heat blanched because of the damage from the heat and the associated sogginess and juice loss after thawing. Instead, chemicals are commonly used without heat to inactivate the oxidative enzymes or to act as antioxidants and they are combined with other treatments.

Ascorbic/Citric Acid Dip

Ascorbic acid or vitamin C minimises fruit oxidation primarily by acting as an antioxidant and itself becoming oxidised in preference to catechol-tannin compounds. Ascorbic acid is frequently used by being dissolved in water, sugar syrup or in citric acid solutions.

It has been found that increased acidity also helps retard oxidative colour changes and so ascorbic acid plus citric acid may be used together. Citric acid further reacts with (chelates) metal ions thus removing these catalysts of oxidation from the system.

Sulphur Dioxide Treatment

Sulphur dioxide may function in several ways:

- sulphur dioxide is an enzyme poison against common oxidising enzymes;
- it also has antioxidant properties; i.e., it is an oxygen acceptor (as is ascorbic acid);
- further SO2 minimises non enzymatic Maillard type browning by reacting with aldehyde groups of sugars so that they are no longer free to combine with amino acids;
- sulphur dioxide also interferes with microbial growth.

In many fruit processing pre-treatments two factors must be considered:

a. sulphur dioxide must be given time to penetrate the fruit tissues;

b. SO_2 must not be used in excess because it has a characteristic unpleasant taste and odour, and international food laws limit the SO_2 content of fruit products, especially of those which are consumer oriented (e.g. except semi-processed products oriented to further industrial utilisation).

Commonly a 0.25 % solution (except for semi-processed fruit products which are industry oriented and use a 6% solution) of SO_2 or its SO_2 equivalent in the form of solutions of sodium sulphite, sodium bisulphite or sodium/potassium metabisulphite are used.

Fruit slices are dipped in the solution for about two to three minutes and then removed so as not to absorb too much SO2. Then the slices are allowed to stand for about one to two hours so that the SO2 may penetrate throughout the tissues before processing.

Sulphur dioxide is also used in fruit juice production to minimise oxidative changes where relatively low heat treatment is employed so as not to damage delicate juice flavour.

Dry sulphuring is the technological step where fruit is exposed to fumes of SO2 from burning sulphur or from compressed gas cylinders;

this treatment could be used in the preparation of fruits (and some vegetables) prior to drying/dehydration.

Sugar Syrup

Sugar syrup addition is one of the oldest methods of minimising oxidation. It was used long before the causative reactions were understood and remains today a common practice for this purpose.

Sugar syrup minimises oxidation by coating the fruit and thereby preventing contact with atmospheric oxygen.

Sugar syrup also offers some protection against loss of volatile fruit esters and it contributes sweet taste to otherwise tart fruits.

It is common today to dissolve ascorbic acid and citric acid in the sugar syrup for added effect or to include sugar syrup after an SO_2 treatment.

Fresh Fruit Storage

Some fruit species and specially apples and pears can be stored in fresh state during cold season in some countries' climatic conditions.

Fruit for fresh storage have to be autumn or winter varieties and be harvested before they are fully mature. This fruit also has to be sound and without any bruising; control and sorting by quality are mandatory operations.

Sorting has to be carried out according to size and weight and also by appearance; fruit which is not up to standard for storage will be used for semi-processed product manufacturing which will be submitted further to industrial processing.

Harvested fruit has to be transported as soon as possible to storage areas. Leaving fruit in bulk in order to generate transpiration is a bad practice as this reduces storage time and accelerates maturation processes during storage.

In order to store large quantities of fruit, silos have to be built.

Technology of Semi-processed Fruit Products

The semi-processed fruit products are manufactured in order to be delivered to industry processing centres (in the fruit producing country itself or in importing countries) where they will be further manufactured in consumer oriented finished products: jams, jellies, syrups, fruits in syrup, etc.

In the practice of semi-processed fruit products and for the purpose of this document the following categories are defined:

a. fruit "pulps": semi-processed products, not refined, obtained by mechanical treatment (or, less often, by thermal treatment) of fruit followed by their preservation. Either whole fruit, halves or big pieces are used which enables easy identification of the species. "Pulps" can be classified in boiled or non boiled (raw).

b. fruit "purées-marks": semi-processed products obtained by thermal and mechanical treatment or, very rare, raw and then refined, operations by which all nonedible parts (cores, peels, etc.) are removed. "Purées-marks" are classified in boiled (the more usual case) and non boiled (raw).

c. semi-processed juices: products obtained by cold pressure or very rare by other treatments (diffusion, extraction, etc.) followed by the preservation.

Technical Processes for Preservation of Semi-processed Fruit Products

Preservation can be achieved by chemical means, by freezing or by pasteurization. The choice of preservation process for each individual case is a function of the semi-processed product type and the shelf life needed.

Chemical preservation. In many countries, in practice, this is carried out with sulphur dioxide, sodium benzoate, formic acid and, on a small scale, with sorbic acid and sorbates.

Preservation with sulphur dioxide is a widespread process because of its advantages: universal antiseptic action and very economic application. The drawbacks of SO_2 are: SO_2 turn firms the texture of some fruit species (pomaces), desulphiting is not always complete and recolouring of red fruits is not always complete after desulphitation.

Practical preservation dosage levels with SO_2 for about 12 months is 0.18-0.20% SO_2 (with respect to the product to be preserved).

This level could be reduced to 0.09% SO_2 for 3 months and to 0.12% SO_2 for 6 months preservation.

The preservation with sulphur dioxide is in use mainly for "pulps" and for "purées-marks".

Chemical preservation can be performed from a practical point of view by the utilisation of 6% SO2 water solutions or by direct introduction of sulphur dioxide gas in the product (for "purées-marks"). The preparation of 6% SO2 solutions is done by bubbling the gas from cylinders in cold water; from a 50 kg SO_2 compressed gas cylinder results 830 l of 6% SO2 solution.

These SO_2 solutions have to be stored in cool places, in closed receptacles and with periodic concentration control/check by titration or by density measurements approximate results.

Preservation with sodium benzoate has the following advantages: it does not firm up the texture and does not modify fruit colour. The disadvantages are: it is not a universal antiseptic, its action needs an acid medium and the removal is partial. Sodium benzoate is in use for "pulps" and for "purées-marks" but less for fruit juices.

Practical dosage level for 12 months preservation is 0.18-0.20 % sodium benzoate, depending on the product to be preserved. Sodium benzoate is used as a solution in warm water; the dissolution water level has to be at maximum 10% reported to semi-processed product weight.

Formic acid preservation is performed mainly for semi-processed fruit juices at a dosage level of 0.2 % pure formic acid (100%). Formic acid is an antiseptic effective against yeasts, does not influence colour of products and is easily removed by boiling.

Formic acid could be diluted with water in order to insure a homogeneous distribution in the product to be preserved; water has to be at maximum 5 % of the product weight. Because of a potential effect of pectic substance degradation, formic acid is less in use for "pulps" and "purées-marks" preservation.

Sorbic acid used as potassium sorbate (easily water soluble) can be used for preservation of fruit semi-processed products at a dosage level of 0. 1% maximum. Advantages of sorbates are: they are completely harmless and without any influence on the organoleptic properties of semi-processed fruit products.

Preservation by pasteurization. As fruit has a low pH, preservation of semiprocessed fruit products could also be performed by pasteurization (heat treatment step at maximum temperature of 100° C), the length of this step varying with the size of the receptacles.

The advantages of this type of treatment are: hygienic process, which assure a long term preservation; the disadvantages are: need for air tight receptacles, and pectic substances could begin to deteriorate if the thermal treatment is too long.

Thermal preservation of fruit semi-processed products could also be done by a "selfpasteurization": very hot semi-processed products are filled into receptacles (e.g. metal cans) which are sealed and then inverted in order to sterilise the air which goes through the hot fruit mass.

Preservation by freezing. This is done on an industrial scale in some countries and can be done with or without sugar addition. The advantages of this process are: absence of added substances; very good preservation of quality of fruit constituents (pectic substances, vitamins, etc.) and good preservation of organoleptic properties (flavour, taste, colour). Freezing is done at about -20 to -30° C and storage at -10 to -18° C.

Freezing is applied mainly to semi-processed fruit products aimed at very high quality and high cost finished products.

Technological flow-sheet for semi-processed fruit "pulps": chemical preservation.

Sorting is needed in order to remove sub-standard fruit (with moulds, with diseases, etc.) and all foreign bodies.

Washing is obligatory in order to remove all impurities which cannot be eliminated at the processing step in finished products.

Coring and Cutting, mainly for pomace fruits, has as main objective a better utilisation of preservation "space" in receptacles and is not mandatory; this will be defined by customer/supplier agreements/ standards. This operation is preferably performed by mechanical means.

Preservation is carried out with the 6% SO2 solution which is added to the prepared fruits (placed in bulk in receptacles) in the quantity needed to obtain the preservation dosage level. For a better/ homogeneous preservative distribution, the initial 6% SO2 solution could be diluted with water; however, the diluted solution (which will be filled in receptacles) has to be at a dosage level of less than 10% of the semi-processed product weight.

For some soft fruit, especially strawberries, preservation is done with a mix of 6% SO2 solution and calcium bisulphite solution (containing also 6% SO2).

Preparation of calcium bisulphite solution is done by the introduction of 30 kg of CaO in 1 m^3 SO2 solution and mixing up to clarification. The resulting solution is mixed with the initial 6% SO2 solution, generally in a 1:1 ratio, but the ratio can be adapted to the fresh fruit texture. Firming of soft fruit texture by this treatment is based on the formation of calcium pectate with pectic substances from fruit tissues.

In the case of sodium benzoate, formic acid or potassium sorbate, the dosage levels to be used are as indicated above with the rule that it is not allowed to add more that 10% liquid in receptacles on the prepared fruits.

Preservation by pasteurization or "self-pasteurization" will need as additional steps: a) boiling with a minimum water addition (maximum 10%); b) filling of receptacles; c) hermetic closing followed by d) pasteurization or "self-pasteurization".

Some general technical data for the preparation of chemical preserved semi-processed fruit "pulps" are seen in Table below.

Table: General technical processing data for semi-processed fruit "pulps"

Fruit species	Preliminary operations	Preservation means
Apples, pears, quinces	Sorting, washing, coring, cutting	Sulphur dioxide
Prunes, peaches, wax cherries, apricots	Sorting, washing, stone removal (pitting)	Sulphur dioxide or sodium benzoate
Cherries	Sorting, washing	Sulphur dioxide or sodium benzoate, sometimes with calcium bisulphite addition
Strawberries	Sorting, washing	Sulphur dioxide in mix with calcium bisulphite
Wild berries	Sorting, washing	Sulphur dioxide or sodium benzoate; in some cases with calcium bisulphite addition

Technological Flow-sheet for Semi-processed "Purée-marks"

The general technological flow-sheet includes the following operations:

Sorting and Washing are obligatory and are carried out in a similar manner as for "pulps".

Heat Treatment/Boiling is needed in order to soften the fruit tissues before refining. For some fruits as strawberry and wild berries, this step is not done and fruits are refined "raw" in order to preserve their flavour.

Pulping is performed with specific equipment - refiners or pulpers - which eliminate seeds, pits and other non edible parts (peels, cores, etc.).

Preservation is carried out by chemical means, by freezing or by pasteurization.

The general technical/processing data for the manufacture of "purée-marks" are seen in Table below.

Table: General technical processing data for semi-processed "purées - marks"

Fruit species	Preliminary operations	Preservation means
Apples, pears, quinces	Sorting, washing, boiling, refining	Sulphur dioxide; in less frequent cases formic acid or sodium benzoate
Prunes, peaches, wax cherries, apricots,	Sorting, washing, boiling, refining	Sulphur dioxide, formic acid or sodium benzoate; cherries freezing with or without sugar; self-pasteurization
Strawberry, wild berries	Sorting, washing, refining	Chemical preservation, freezing or self pasteurization

Fruit Sugar Preserves Technology; Jams, Jellies, Marmalade, Fruit Paste

As a overall rule of thumb, a sugar concentration of about 60% in finished or processed fruit products generally insures their preservation. Preservation is not only determined by the osmotic pressure of sugar solutions but also by the water activity values in the liquid phase, which can be lowered by sugar addition; and by evaporation down to 0.848 aw; this value however does not protect products from mould and osmophile yeast attack.

Maximum saccharose concentration that can be achieved in the liquid phase of the product is 67.89%; however higher total sugar quantities (up to 70-72%) found in products are explained by an increased reducing sugar solubility resulting from saccharose inversion.

Jams

The preservation of fruit by jam making is a familiar process carried out on a small scale by housewives in many parts of the world. Factory jam making has become a highly complex operation, where strict quality control procedures are employed to ensure a uniform product, but the manufacturing operations employed are in essence the same as those employed in the house.

Fresh or pre-cooked fruit is boiled with a solution of cane or beet sugar until sufficient water has been evaporated to give a mixture which will set to a gel on cooling and which contains 32-34% water.

Gel formation is dependent on the presence in the fruit of the carbohydrate pectin, which at a pH of 3.2 - 3.4 and in the presence of a high concentration of sugar, has the property of forming a viscous semi-solid. During jam boiling, all micro-organisms are destroyed within the product, and if it is filled hot into clean receptacles which are subsequently sealed, and then inverted so that the hot jam contacts the lid surface, spoilage by micro-organisms will not take place during storage.

The composition of jam made from stone fruit and berry fruit is shown in Table below. About 30% of the vitamin C present in fresh fruit is destroyed during the jam-making process, but that which remains in the finished product is stable during storage.

The high moisture content of jam (equivalent to an equilibrium relative humidity of about 82%) makes it susceptible to mould damage once the receptacle has been opened and exposed from some time to the air. No problems of microbiological spoilage are likely to arise in the canned product during storage.

Table: Composition of some fruit jams

Type of Jam	Moisture content %	Sugar (as invert sugar, %)	Vitamin C mg/100 g
Jam made from berry fruits: strawberry, raspberry, etc.	29.8	69.0	10 - 25
Jam made from stone fruits: apricot, peach, etc.	29.6	69.3	10 - 35

Source: FAD/WFP, 1970

Marmalade

This sugar preserve is defined as "semisolid or gel-like product prepared from fruit ingredients together with one or more sweetening ingredients and may contains suitable food acids and food pectins; the ingredients are concentrated by cooking to such a point that the TSS - Total Soluble Solids - of the finished marmalade is not below 65%".

Fruit Paste

Fruit paste is a product obtained in the same way as special non-gelified fruit marmalade but with a lower water content - about 25% TSS in fruit paste.

Lowering water content could be achieved by continuing boiling of the product or by drying the product by natural or artificial drying. An example of paste without sugar is the sun dried apricot or prune paste.

General procedure for the preparation of jams, jellies and marmalade:

a. Boil the pulp or the juice (with water when necessary)
b. Add the pectin
    * to the batch while stirring very vigorously
    * Pectin which has previously been mixed with 5 times its weight in sugar taken from the recipe)
c. Boil for about 2 minutes to assure a complete dissolution
d. Add the sugar while keeping the batch boiling
e. Boil down quickly to desired Brix
f. Add the acid (usually citric acid) and remove the froth
g. Fill hot into the (previously cleaned) jars and close
h. Invert the jars for three minutes to pasteurize the cover.

* Note: the pectin in solution can also be added at the end of the step (e) and has to be prepared as follows: use a strong blender.

For one litre of water add slowly into the blender 25 g of pectin mixed with 100 g of sugar taken from the recipe.

Basic recipes

The following recipes must be considered only as guidelines because the composition of the fruit can vary; also the taste of the consumers varies concerning the consistency, the sweetness and acidity.

Before starting to make jam it is important to know the yield to settle the question on containers.

The calculation is made as follows:

$$\frac{\text{Total Soluble Solids content of all the raw ingredients}}{\text{Percentage of Total Soluble Solids in the final product}} \times 100$$

In these basic recipes it is assumed that the fruits are poor in pectin content.

Recipe 1. Fruit: sugar = 50:50; desired Brix in the finished product is 68.

Soluble Solids:

10 kg of fruit at 10% TSS	1.000 kg
10 kg of sugar	10.000 kg
60 g of pectin (grade 200)	0.060 kg
55 g of citric acid	0.055 kg
	11.115 kg

$$\text{Yield} = \frac{11.115 \times 100}{68} = 16.4 \text{ Kg}$$

Recipe 2. Fruit: sugar = 45:55; desired Brix in the finished product is 68.

Soluble Solids

d10 kg of fruit at 10% TSS	1.000 kg
2.5 litre of water	-
12.2 kg of sugar	12.200 kg
65 g of pectin grade 200	0.060 kg
60 g of citric acid	0.060 kg
	13.325 kg

$$\text{Yield} = \frac{13.325 \times 100}{68} = 19.6 \text{ Kg}$$

Recipe 3. Fruit: sugar = 40:60; desired Brix in the finished product is 68.

Soluble Solids

d10 kg of fruit at 10% TSS	1.000 kg
3.3 litre of water	-
15 kg of sugar	15.000 kg
85 g of pectin grade 200	0.085 kg
80 g of citric acid	0.080 kg
	16.165 kg

$$\text{Yield} = \frac{16.165 \times 100}{68} = 23.8 \text{ Kg}$$

Various factors must be taken into account:

1. Size of the container: the quantity of pectin indicated in the recipes is valid for containers of 1 kg or less.

If container capacity is between:	*Increase pectin by:*
1 kg and 2.5 kg	5%
2.5 kg and 5.0 kg	10%
5.0 kg and 10.0 kg	20%
10.0 kg and 20.0 kg	30%

2. Finishing point: the quantity of pectin is given for a final Brix - Total Soluble Solids (TSS) of 68%.

   If the final Brix is 66 increase the pectin by 5%

   " " is 65 " " " by 10%

   is 64 " " " by 15%

   " " is 62 " " " by 20%

   " " is 60 " " " by 30%
3. Acidic taste. If the product is too acid, replace the citric acid by tartaric acid (63% of the amount of citric acid).
4. Formation of clots: If batch clots, it is probably due to the pH being too low or the or TSS being too high; correct accordingly.
5. Formation of liquid at the surface: if liquid forms on the surface, it is probably due to too low a pH or too low pectin content.
6. Crystallisation: a) if liquid forms on the surfaces, then the pH is too low; reduce the acid content; b) if liquid does not form on the surface, then TSS or pH is too high.
7. Formation of mould: the TSS is probably below 68 deg. Brix. The filling may have been done at a low temperature. If the containers are large, wait until they are cold before closing.
8. Wrong batch: dilute the jam with water to 30% TSS; cook briefly. Add this diluted jam to a new batch but in a ratio not exceeding 10%.

Processing of Pineapple-papaya Jam

The fruit should be prepared as per previous instructions. For pineapples, the ends are removed and discarded; the cores and outer parts of the fruits are also removed. The fruit cylinders obtained are pulped through a special extractor (Fitzpatrick communiting machine) equipped with a 0.40-in screen sieve; the pulp thus obtained is used for making jam.

The papaya are prepared by hand-peeling the fruit; the fruit is then halved and the seeds removed. It is then pulped in the communiting machine using a 0.40-in screen sieve. When ginger root is used as flavouring, it is peeled and macerated in a Kenwood blender to a very fine consistency.

A typical formula for a pineapple-papaya jam (50:50 ratio) with ginger flavouring is given as follows:

Pineapple pulp	25.0
Papaya pulp	25.0 pounds
Cane sugar	50.0
Apple pectin (150 grade)	6.0 ounces
Citric acid	6.4
Fresh ground ginger	7.5

Processing is carried out in the following way:

The weighed fruit pulp is placed in a stainless steel steam-jacketed kettle and heated to about 110°F under constant stirring.

When the product reaches this temperature, the heat is turned off. The pectin (mixed in about ten times its weight with some of the weighed sugar), is then mixed into the fruit pulp, stirring constantly in order to prevent the pectin from clotting.

When the pectin has dissolved, the remainder of the sugar is added and dissolved completely in the mixture. The heat is then turned on and the jam mixture is stirred constantly until it starts boiling vigorously. During the remainder of the cooking, the product is stirred occasionally. Near the finishing point (approximately 221° F), the citric acid and the ginger (if it is used) are also added.

Determination of the finishing point is done by removing samples at intervals, cooling, and reading the soluble solids by means of a refractometer equipped with a Brix scale. After the jam reaches the proper Total Soluble Solids content, the heat is turned off and the surface scum/foam is removed.

The jam then is quickly put into receptacles which have been cleaned and sterilised with boiling in water for 30 minutes. The filling operation is done rapidly in order to prevent the temperature of the jam from falling below 190° F.

After filling, sterilised lids (boiled for 30 minutes in water) are placed on the receptacles and they are then sealed.

After this operation the receptacles are inverted for about 3 minutes to insure that the lids are sterilised. The receptacles are then placed upright. At this stage it is not necessary to do any further processing, therefore the receptacles are cooled in running cold water until they reach a temperature slightly above room temperature. They are then dried in air and labelled.

Evaluation of Finished Products

During production at medium/large scale, it is recommended that quality controls be performed during manufacturing. After ten weeks of storage at room temperature it is recommended that an examination of finished products be performed. The receptacles are opened and contents carefully emptied on to enamel trays without disturbing the formation of the jam.

The empty cans (if metal cans were used) are then inspected for signs of corrosion. Factors other than flavour include colour, appearance, syrup separation, firmness and spreading quality. For flavour, jam is tested on pieces of bread. Samples are taken for measurement of pH (with a glass electrode pH meter) and Total Soluble Solids (with a refractometer equipped with a Brix scale).

This evaluation enables to have a quality check during product shelf life and to obtain data needed for necessary improvements of future productions.

For pineapple-papaya jam, products made with 30% pineapple and 70% papaya with added ginger has the highest score for flavour. The use of plain tin cans causes corrosion problems which is not the case when acid resistant lacquer cans are used.

Pineapple Jam Making

1. Boil 40 lb. of pulp and 12 lb. of water.
2. Add 225 g of pectin to the batch while stirring rapidly.
3. Boil for about 90 sec to assure complete dissolution.
4. Add 60 lb. of sugar gradually if possible in several portions, while keeping the batch boiling.
5. Boil down quickly to 69 deg. Brix (223 deg. F).
6. Take off steam, remove foam.
7. Add 300 cc citric acid solution 50%.

8. Fill hot (180 deg. F).
9. Invert receptacles for 3 minutes.

Check each batch for Brix 68-70 deg.; acidity/pH = 3.2 +/- 0.2.

Evaluate: absence of defects; colour; flavour; consistency.

Various fruit: sugar ratios can be manufactured; some basic recipes are as follows:

	Ratio 50:50	*Ratio 45:55*	*Ratio 40:60*
Fruit	55 lb	49.5 lb	44 lb
Water	-	11 lb	13.2 lb
Sugar	55 lb	60.5 lb	66 lb
Pectin (150 grade)	225 g	237.5 g	250 g
Citric acid (50% sol.)	300 cc	320 cc	335 cc

Fruit Juice Technologies

Fruit juices are products for direct consumption and are obtained by the extraction of cellular juice from fruit, this operation can be done by pressing or by diffusion.

For the purpose of this document, the technology of fruit juice processing will cover two finished product categories:

- juices without pulp ("clarified" or "not clarified");
- juices with pulp ("nectars").

We will also define as

a. "natural juices" products obtained from one fruit; and
b. "mixed juices" products obtained from the mix of two or three juices from different fruit species or by adding sugar.

Juices obtained by removal of a major part of their water content by vacuum evaporation or fractional freezing will be defined as "concentrated juices".

Technological Steps for Processing of Fruit Juices Without Pulp

Fruit juices must be prepared from sound, mature fruit only.

Soft fruit varieties such as grapes, tomatoes and peaches should only be transported in clean boxes which are free from mould and bits of rotten fruit.

Washing: fruit must be thoroughly washed. Generally, fruit will be submitted to a pre-washing before sorting and a washing step just after sorting.

Sorting: removal of partially or completely decayed fruit is the most important operation in the preparation of fruit for production of first quality fruit juices; sorting is carried out on moving inspection belts or sorting tables.

Crushing/Grinding/Disintegration Step is applied in different ways and depends on fruit types:

Crushing for grapes and berries;

Grinding for apples, pears;

Disintegration for tomatoes, peaches, mangoes, apricots.

This processing step will need specific equipment which differs from one type of operation to another.

Enzyme Treatment of crushed fruit mass is applied to some fruits by adding 2-8% pectolitic enzymes at about 50° C for 30 minutes.

This optional step has the following advantages: extraction yield will be improved, the juice colour is better fixed and finished product taste is improved.

However, for fruit which is naturally rich in pectic substances, this treatment makes the resulting "exhausted" material useless for industrial pectin production.

Heating of crushed fruit mass before juice extraction is an optional step used for some fruit in order to facilitate pressing and colour fixing; at same time, protein coagulation takes place.

Pressing to extract juice.

Diffusion is an alternative step for juice extraction and can be carried out discontinuously or in batteries at water temperature of about 80-85 ° C.

Juice Clarifying can be performed by centrifugation or by enzyme treatment. Centrifugation achieves a separation of particles in suspension in the juice and can be considered as a pre-clarifying step. This operation is carried out in centrifugal separators with a speed of 6000 to 6500 RPM.

Enzyme clarifying is based on pectic substance hydrolysis; this will decrease the juices' viscosity and facilitate their filtration. The treatment is the addition of pectolitic enzyme preparations in a quantity of 0.5 to 2 g/l and will last 2 to 6 hours at room temperature, or less than 2 hours at 50° C, a temperature that must not be exceeded.

The control of this operation is done by checking the decrease in juice viscosity. Sometimes, the enzyme clarifying is completed with the step called "sticking" by the addition of 5-8 g/hl of food grade gelatine which generates a flocculation of particles in suspension by the action of tannins.

Filtration of clarified juice can be carried out with kieselgur and bentonite as filtration additive in press-filters (equipment).

De-tartarisation is applied only to raisin juice and is aimed to eliminate potassium bitartrate from solution. This step can be performed by the addition of 1% calcium lactate or 0.4% calcium carbonate.

Pasteurization of juice can be done for temporary preservation (pre-pasteurization) and in this case this operation is carried out with continuous equipment (heat exchangers, etc.); warm juice is stored in drums or large size receptacles (20-30 kg). Pasteurization conditions are at 75°C in continuous stream.

Pasteurization of bottled juice is then carried out just before delivery to the market; this is performed in water baths at 75° C until the point where the juice reaches 68° C. In cases when the final pasteurization is done without pre-pasteurization and temporary storage, modern methods use a rapid pasteurization followed by aseptic filling in receptacles.

Rapid pasteurization conditions are as follows: temperature about 80° C, over 10-60 sec., followed by cooling; all operations are carried out in continuous stream.

Preservation under CO2 pressure may be done at a concentration of 1.5% CO2 under a pressure of 7 kg/cm^2. At the distribution step, proceed at CO2 decompression and the juice is then submitted to a sterilising filtration and aseptic filling in receptacles.

Preservation by freezing is carried out at about -30° C, after a preliminary de-aeration; storage is at -15 to -20° C.

Production of concentrated juices by evaporation is performed under vacuum (less than 100 mm Hg residual pressure) up to a concentration of 65-70% total sugar which assures preservation without further pasteurization. Modern evaporation installations recover flavours from juices which are then reincorporated in concentrated juices.

Additional operations for juice manufacturing are the vacuum de-aeration and mixing with other fruit juices or with sugar.

For the production of non clarified juices the centrifugation is the only specific step, enzyme clarifying and subsequent filtration being eliminated.

The optimum sugar/acid ratio for the majority of fruit, mainly for pomaces, is 10/1 to 15/1.

Fruit which is rich in carotenoids (apricots, peaches, etc.) is only processed as juices with pulp ("nectars").

Technological Flow-sheet for Fruit Juices with Pulp ("Nectars")

This process is divided at industrial scale in two categories of operations:

a. processing for obtaining juices;
b. juice conditioning for preservation.

Operations in the first category are differ according to the type of fruit which to be processed. Pomaces (apples, pears) are washed and sorted and then crushed in a colloid mill; fruit purée is then passed through a screw type heating equipment where direct steam is used as a source of heat. Warm fruit mass is treated in a pulper with a 2 mm screen and then through an extractor similar with the equipment used for tomato juice.

Stone fruits (apricots, peaches, cherries, etc.) after washing and sorting are submitted to steam in a continuous heater, then the warm fruit mass is passed through a pulper and then an extractor (as mentioned above). Berries (strawberry, wild berries, etc.) are washed, sorted and then crushed, preheated and then introduced in extractor.

In order to avoid browning and undesirable taste modifications it is usual to add about 0.05% ascorbic acid.

Second category type of operations are similar for all fruit species. Partial elimination of cellulose is achieved with a continuous centrifugal separator; the resulting juice is then processed in order to adjust sugar and acid content for viscosity.

Sugar (about 8-10%) is added as a syrup (in water or in the juice of same fruit obtained by pressure). Acidity is adjusted with citric or tartaric acid. The adjusted juice is then deaerated under vacuum at about 40° C. This step aims at avoiding oxidative reactions and vitamin C loss reduction.

An important subsequent step is an intensive homogenisation (under pressure at 150-180 A) in order to obtain particles with dimensions below 100. The homogenised juice obtained is then continuously pasteurized in plate heat exchanger equipment at a temperature of about 130° C, cooled down to about 90° C and aseptically packed in receptacles.

The principal characteristics of fruit "nectars" are uniformity and stability of the content provided by the advanced disintegration of fruits. Stability can be obtained by increasing product viscosity by adding pectin for fruit which is deficient in this component. In order to avoid "separation", intensive homogenisation is carried out as described above.

Fruit "nectars" contain all the important components of the original fruit and to a large extent maintain their taste and flavour. The sugar/ acidity (as citric acid) ratio is to a large extent determined by the type of fruit and the correction applied; for example, this ratio is 30 for apricots, 40 for peaches, 160 for pears, etc.

Banana and Plantain Processing Technologies

Products: Uses and Dietary Significance

Most of the world's bananas are eaten either raw, in the ripe state, or as a cooked vegetable, and only a very small proportion are processed in order to obtain a storable product. This is true both at a traditional village level with both dessert and cooking bananas and when considering the international trade in dessert bananas.

In general, preserved products do not contribute significantly to the diet; however, in some localised areas the products are important in periods when food are scarce.

Probably the most widespread and important product is flour preparation from unripe banana and plantains by sun-drying. In Uganda, dried slices known as "mutere" are prepared for storage from green bananas, the dried slices being either used directly for cooking or after grinding into a flour. "Mutere" is used chiefly as a famine reserve and does not feature largely in the diet under normal conditions.

In Gabon, plantains are sometimes made into dried slices which can be stored and used on long journeys, and plantains are used in Cameroon to prepare dried pieces which are stored and ground as needed into flour for use in cooking a paste known as "fufu". Dried green banana slices are also used in parts of South and Central America and West Indies for preparing flour. The other nutritionally important product is beer which is a major product in Uganda, Rwanda and Burundi where green banana utilisation is particularly high.

Preservation Methods and Processes

Drying. - Both ripe and unripe bananas and plantains are normally peeled and sliced before drying, although banana figs are sometimes prepared from whole ripe fruit. Sun drying is the most widespread technique where the climate is suitable but drying in ovens or over fires is also practiced. In west Africa, plantains are often soaked and sometimes parboiled before drying. The slices of unripe fruit are normally spread out on bamboo frameworks; or a cemented area; or on a mat; or on a swept-bare patch of earth; or on a roof; or sometimes on stones outcrops or sheets of corrugated iron. Oven-drying of ripe bananas is practiced in Polynesia as a mean of preserving the fruits, which are then wrapped in leaves and bound tightly to store until needed. In East Africa a method has been reported that involves drying the peeled bananas on a frame placed over a fire for 24 hr before drying in the sun, to accelerate the process.

Potential for Scaling up of Traditional Processes to Industrial Level

Many banana products are now produced on an industrial scale, including the traditional banana figs and flour, and the processing techniques are described below. One of the main problems encountered has been the susceptibility of banana products to flavour deterioration and discoloration and in the past many products reaching the market have been of poor quality.

A great deal of research has been directed to overcoming these problems, although however good the resultant products are they cannot compare in flavour and other characteristics with the fresh banana fruit. Indeed, an important constraint on the large-scale development of banana processing is the lack of demand for banana products since the fresh fruit is available throughout the year in most parts of the tropical world.

The production of beer from banana and plantains has not been scaled up to an industrial level, and while an important product in localised areas of tropical Africa, the market is rapidly declining in favour of European-type brews produced locally.

Industrial Processing

The main commercial products made from bananas are canned or frozen purée, dried figs, banana powder, flour, flakes, chips (crisps), canned slices and jams.

Banana products can be divided roughly into two types - those for direct consumption, such as figs, and those for use in food manufacturing industry, for example purées and powder.

Banana figs, or fingers as they are sometimes known, are usually whole, peeled fruit carefully dried so as to retain their shape, although sometimes the fruit is sliced or halved to facilitate drying. Banana and plantain chips (crisps) are thinly sliced pieces of fruit fried in oil and eaten as a snack like potato chips (crisps).

The main use of canned slices is in tropical fruit salads. Banana flakes are used as a flavouring or in breakfast cereals. Banana purée find use mainly in the production of baby foods. Banana flour is said to be highly digestible and is used in baby and invalid foods, but can also be used in the preparation of bread and beverages.

Banana powder is used chiefly in the baking industry for the preparation and fillings for cakes and biscuits and is also used for invalid and baby foods.

Processing Technology

In general, to obtain a good-quality product from ripe-bananas the fruit is harvested green and ripened artificially under controlled conditions at the processing factory. After ripening, the banana hands are washed to remove dirt and any spray residues, and peeled. Peeling is almost always done by hand using stainless steel knives, although a mechanical peeler for ripe bananas has been developed, capable of peeling 450 Kg of fruit per hour (Banana Bulletin, 1974).

The peeling of unripe bananas and plantains is facilitated by immersing the fruit in hot water. For example, immersion in water at 70-75 ° C for 5 min. has been suggested as an aid for peeling green bananas for flour production, while the peeling of green bananas for freezing has been facilitated by immersion in water at 93° C for 30 min.

Banana Figs

Fully ripe fruits with a sugar content of about 19.5% are used and are treated with sulphurous acid after peeling, then dried as soon as possible afterwards. Various drying systems have been described using temperatures between 50 and 82° C for 10 to 24 hr to give a moisture content ranging from 8 to 18% and a yield of dried figs of 12 to 17% of the fresh banana on the stem.

One factory in Australia uses a solar heat collector on the roof to augment the heat used for drying bananas. Bananas can also be dried by osmotic dehydration, using a technique which involves drawing water from 1/4-in. thick banana by placing them in a sugar solution of 67 to 70 deg. Brix for 8 to 10 hr. followed by vacuum-drying at 65 to 70° C, at a vacuum of 10 mm Hg for 5 hr. The moisture content of the final products is 2.5% or less, much lower than that achieved by other methods.

Banana Purée

Banana purée is obtained by pulping peeled, ripe bananas and then preserving the pulp by one of three methods: canning aseptically, acidification followed by normal canning, or quick-freezing.

The bulk of the world's purée is processed by the aseptic canning technique. Peeled, ripe fruits are conveyed to a pump which forces them through a plate with 1/4-in. holes, then onto a homogeniser, followed by a centrifugal de-aerator, and into a receiving tank with 29in. vacuum, where the removal of air helps prevent discoloration by oxidation.

The purée is then passed through a series of scraped surface heat exchangers where it is sterilised by steam, partially cooled, and finally brought to filling temperature. The sterilised purée is then packed aseptically into steam-sterilised cans which are closed in a steam atmosphere.

Banana Slices

Several methods for canning of banana slices in syrup are used. Best-quality slices are obtained from fruit at an early stage of ripeness. The slices are processed in a syrup of 25 deg. Brix with pH about 4.2, and in some processes calcium chloride (0.2%) or calcium lactate (0.5%) are added as firming agents.

A method for producing an intermediate-moisture banana product for sale in flexible laminate pouches has been developed. Banana slices are blanched and equilibrated in a solution containing glycerol (42.5%), sucrose (14.85%), potassium sorbate (0.45%), and potassium metabisulphite (0.2%) at 90 deg. C for 3 min. to give a moisture content of 30.2%.

Banana Powder

In the manufacture of banana powder, fully ripe banana pulp is converted into a paste by passing through a chopper followed by a colloid mill. A 1 or 2 % sodium metabisulphite solution is added to improve the colour of the final product. Spray- or drum-drying may be used, the latter being favoured as all the solids are recovered.

A typical spray dryer can produce 70 kg powder per hour to give yields of 8 to 11% of the fresh fruit, while drum-drying gives a final yield of about 13% of the fresh fruit. In the latter method the moisture content is reduced to 8 to 12 % and then further decreased to 2 % by drying in a tunnel or cabinet dryer at 60° C.

Banana Flour

Production of flour has been carried out by peeling and slicing green fruit, exposure to sulphur dioxide gas, then drying in a counter-current tunnel dryer for 7 to 8 hr. with an inlet temperature of 75° C and outlet temperature of 45° C, to a moisture content of 8%, and finally milling.

Banana Chips (Crisps)

Typically, unripe peeled bananas are thinly sliced, immersed in a sodium or potassium metabisulphite solution, fried in hydrogenated oil at 180 to 200° C, and dusted with salt and an antioxidant.

Alternatively, slices may be dried before frying and the antioxidant and salt added with the oil. Similar processes for producing plantain chips have been developed.

Banana Beverages

In a typical process, peeled ripe fruit is cut into pieces, blanched for 2 min. in steam, pulped and pectolytic enzyme added at a concentration of 2 g enzyme per 1 kg pulp, then held at 60 to 65° C and 2.7 to 5.5 pH for 30 min.

In a simpler method, lime is used to eliminate the pectin. Calcium oxide (0.5%) is added to the pulp and after standing for 15 min. this is neutralised giving a yield of up to 88% of a clear, attractive juice. In another process banana pulp is acidified, and steam-blanched in a 28-in Hg vacuum which ensures disintegration and enzyme inactivation.

The pulp is then conveyed to a screw press, the resulting purée diluted in the ratio 1:3 with water, and the pH adjusted by further addition of citric acid to 4.2 to 4.3, which yields an attractive drink when this is centrifuged and sweetened.

Jam

A small amount of jam is made commercially by boiling equal quantities of fruit and sugar together with water and lemon juice, lime juice or citric acid, until setting point is reached.

Product Stability and Spoilage Problems

All dried banana products are very hydroscopic and susceptible to flavour deterioration and discoloration, but this can be overcome to some extent by storing in moisture-proof containers and sulphiting the fruit before drying to inactivate the oxidases.

The dried products are also liable to attack by insects and moulds if not stored in dry conditions, although disinfestation after drying by heating for 1 hr to 80° C or by fumigation with methyl bromide ensures protection against attack. Banana powder is said to be stored for up to a year commercially and flakes have been stored in vacuum-sealed cans with no deterioration in moisture, colour or flavour for 12 months.

Banana chips tend to have a poor storage life and to become soft and rancid. However, chips treated with an antioxidant have been stored satisfactorily at room temperature in hermetically sealed containers up to 6 months with no development of rancidity.

Quality Control Methods

In general a good quality product is obtained if fruit is harvested at the correct stage of maturity and, where appropriate, ripened under controlled conditions. For example, in the case of banana figs, the fruit should be fully mature (sugar content of 19.5% or above) or the final alternatively, slices may be dried before frying and the antioxidant and salt added with the oil. Similar processes for producing plantain chips have been developed.

Preparation Methods for Fresh Bananas and Plantains

The main ways of preparing fresh bananas for consumption are boiling or steaming, roasting or baking and frying. Boiling followed by pounding into "fufu" is also widely adopted in certain areas of the tropics.

Boiling or Steaming

Plantains and bananas are often prepared simply by boiling in water, either in their peel or after peeling, and either ripe or unripe; if unripe, the fruit is scraped thoroughly after peeling to remove all traces of fibrous material. The boiled fruit is eaten alone or more usually accompanied by a sauce. This preparation technique is widely used in West Africa.

Roasting or baking. Unpeeled or peeled fruit, either ripe or unripe, is roasted simply by placing in the ashes of a fire or in an oven. This method is widely used in West Africa, East Africa and the South Pacific islands. For example, ripe plantains are placed unpeeled in an oven and when partly brown and tender, removed and peeled, then replaced in the oven and roasted evenly.

Frying. Ripe or unripe plantains or bananas are often peeled, sliced and cooked in oil, particularly in West Africa and in parts of South America and the West Indies. Similar products are also made in East Africa. Typically, ripe plantains are peeled, cut into slices or split lengthways, and fried in palm oil or with groundnut oil, the pieces being served either hot with a sauce or with fried eggs, or cold as a snack.

Pounding. Pounding is a process, used particularly in West Africa, for preparing most perishable staple food crops including plantains, cassava, yams and cocoyams to obtain a paste or dough known as "fufu" (also spelled "foofoo", "foutou", "foufou"). The plantains are peeled or boiled and peeled after boiling and pounded in a wooden mortar, the resulting paste normally being eaten with soup or a spiced sauce of meat and vegetables, but sometimes after wrapping in leaves and steaming.

Mango and Guava Processing Technologies

Mangoes are processed at two stages of maturity. Green fruit is used to make chutney, pickles, curries and dehydrated products. The green fruit should be freshly picked from the tree. Fruit that is bruised, damaged, or that has prematurely fallen to the ground should not be used. Ripe mangoes are processed as canned and frozen slices, purée, juices, nectar and various dried products. Mangoes are processed into many other products for home use and by cottage industry.

The mango processing presents many problems as far as industrialization and market expansion is concerned. The trees are alternate bearing and the fruit has a short storage life; these factors make it difficult to process the crop in a continuous and regular way. The large number of varieties with their various attributes and deficiencies affects the quality and uniformity of processed products.

The lack of simple, reliable methods for determining the stage of maturity of varieties for processing also affects the quality of the finished products. Many of the processed products require peeled or peeled and sliced fruit. The lack of mechanised equipment for the peeling of ripe mangoes is a serious bottleneck for increasing the production of these products.

A significant problem in developing mechanised equipment is the large number of varieties available and their different sizes and shapes. The cost of processed mango products is also too expensive for the general population in the areas where most mangoes are grown. There is, however, a considerable export potential to developed countries but in these countries the processed mango products must compete with established processed fruits of high quality and relatively low cost.

Green Mango Processing

Pickles. The optimum stage of maturity should be determined for each variety used to make pickles.

There are two classifications of pickles - salt pickles and oil pickles. They are processed from whole and sliced fruit with and without stones. Salt is used in most pickles.

The many kinds of pickles vary mainly in the proportions and kinds of spices used in their preparation. One basic recipe for the study of the preparation and storage of pickles in oil is as follows:

Mango pieces	250 g	Tumeric powder	2 to 4 g
Salt	60 g	Fenugreek seeds	2 to 4 g
Mustard powder	30 g	Bengal gram seeds	2 to 4 g
Chili powder	20 g	Gingelly oil	20 to 30 g

The ingredients are mixed together and filled into wide-mouthed bottles of 0.5 kg capacity. Three days later the contents are thoroughly mixed and refilled into the bottles. Extra oil is added to form a 1-2 cm layer over the pickles.

Chutney. The product is prepared from peeled, sliced or grated unripe or semi-ripe fruit by cooking the shredded fruit with salt over medium heat for 5 to 7 minutes, mixed and then sugar, spices and vinegar are added. Cook over moderate heat until the product resembles a thick purée, add remaining ingredients and simmer another 5 min. Cool and preserve in sterilised jars.

Spices usually include cumin seeds, ground cloves, cinnamon, chili powder, ginger and nutmeg. Other ingredients such as dried fruits, onions, garlic and nuts may be added.

Drying/dehydration. Immature fruit is peeled and sliced for sun-drying. The dried mango slices can be powdered to make a product called amchoo. The use of blanching, sulphuring and mechanical dehydration gives a product with better colour, nutrition, storability and fewer microbiological problems.

Ripe Mango Processing

Purée. Mangoes are processed into purée for re-manufacturing into products such as nectar, juice, squash, jam, jelly and dehydrated products. The purée can be preserved by chemical means, or frozen, or canned and stored in barrels. This allows a supply of raw materials during the remainder of the year when fresh mangoes are not available.

It also provides a more economical means of storage compared with the cost of storing the finished products, except for those which are dehydrated, and provides for more orderly processing during peak availability of fresh mangoes.

Mangoes can be processed into purée from whole or peeled fruit. Because of the time and cost of peeling, this step is best avoided but with some varieties it may be necessary to avoid off-flavours which may be present in the skin. The most common way of removing the skin is hand-peeling with knives but this is time-consuming and expensive. Steam and lyepeeling have been accomplished for some varieties.

Several methods have been devised to remove the pulp from the fresh ripe mangoes without hand-peeling. A simplified method is as follows: the whole mangoes were exposed to atmospheric steam for 2 to 2 1/2 min in a loosely covered chamber, then transferred to a stainless steel tank.

The steam-softened skins allowed the fruit to be pulped by a power stirrer fitted with a saw-toothed propeller blade mounted 12.7 to 15.2 cm below a regular propeller blade. The pulp is removed from the seeds by a continuous centrifuge designed for use in passion fruit extraction. The pulp material is then passed through a paddle pulper fitted with a 0.084 cm screen to remove fibre and small pieces of pulp.

Mango purée can be frozen, canned or stored in barrels for later processing. In all these cases, heating is necessary to preserve the quality of the mango purée. In one process, purée is pumped through a plate heat exchanger and heated to 90°C for 1 min and coolcd to 35° C before being filled into 30 lb tins with polyethylene liners and frozen at -23.50 C.

In an other process, pulp is acidified to pH 3.5, pasteurized at 90°C, and hot-filled into 6 kg high-density bulk polyethylene containers that have been previously sterilised with boiling water. The containers are then sealed and cooled in water. This makes it possible to avoid the high cost of cans.

Wooden barrels may be used to store mango pulp in the manufacture of jams and squashes. The pulp is acidified with 0.5 to 1.0% citric acid, heated to boiling, cooled, and SO2 is added at a level of 1000 to 1500 ppm in the pulp. The pulp is then filled into barrels for future use.

Slices. Mango slices can be preserved by canning or freezing, and recent studies have shown the feasibility of pasteurized-refrigerated and dehydro-canned slices. The quality of the processed product in all of these procedures will be dependent upon selection of a suitable variety along with good processing procedures. Thermal process canning of mango slices in syrup is the most widely used preservation method.

Beverages : The commercial beverages are juice, nectar and squash. Mango nectar and juice contain mango purée, sugar, water and citric acid in various proportions depending on local taste, government standards of identity, pH control, and fruit composition of the variety used. Mango squash in addition to the above may contain SO2 or sodium benzoate as a preservative. Other food grade additives such as ascorbic acid, food colouring, or thickeners may be used in mango beverages.

A short description of finished products found in literature is as follows:

- mango juice: prepared by mixing equal quantities of pulp (purée) and water together and adjusting the total soluble solids (TSS) and acidity to taste (12 to 15% TSS and 0.4 to 0.5% acidity as citric acid);
- mango nectar containing 25% purée can be prepared using the following procedure.

		Brix of purée	
Nectar components	15°	17°	20°
Purée	100	100	100
Sugar	45	43	40
Water	255	257	260

Commercial processing conditions may require the use of a preservative.

The pH is adjusted to approximately 3.5 by adding citric acid as a 50% solution.

The time of heat processing will vary with filling temperature, can size and viscosity of the juice or nectar. Mango squash may be prepared according to flow-sheet described below; the finished product may contain 25% juice, 45% TSS and 1.2 to 1.5% acidity and may be preserved with sulphur dioxide (350ppm) or sodium benzoate (1000 ppm) in glass bottles.

Mango squash simplified flow-sheet.

Ingredients

Mango pulp	900	900
Sugar	900	1100
Citric acid	18	15
Water	900	900

Mangoes are washed, stored, peeled with stainless steel knives. The pulp is prepared by using a pulper with fine sieve (0.025-in); Sugar is mixed with water and citric acid = syrup; The pulp is added to the syrup and mixed well; The mixture is strained trough cloth; The squash is heated at 85° C and bottles are filled and closed.

For additional heat treatment bottles may need to be maintained at a product temperature of 80°C for 30 minutes if the product is to be processed without preservatives. The bottles are then left to cool in water and stored at room temperature.

Two negative points must be avoided: presence of air bubbles (which is a source of quick deterioration) and separation of squash solids (giving an undesirable appearance). The means to avoid these two phenomena are described in the fruit juices section.

A type of "squash type" beverage may also be manufactured with '/a pulp + '/a water + i/a sugar and pH adjusted to 3.7 by addition of citric acid. Using different sieve sizes affects the quality and reduces air bubbles to a certain extent but homogenisation and de-aeration of purée or squash seem to be important in order to avoid separation and air bubbles.

The squash quality is evaluated on the basis of the following characteristics: pH, titrable acidity, soluble solids, ascorbic acid (by 2,6 dichlorophenol indophenol method), specific gravity.

Dried/dehydrated. Ripe mangoes are dried in the form of pieces, powders, and flakes. Drying procedures such as sun-drying, tunnel dehydration, vacuum-drying, osmotic dehydration may be used. Packaged and stored properly, dried mango products are stable and nutritious.

One described process involves as pre-treatment dipping mango slices for 18 hr (ratio 1:1) in a solution containing 40° Brix sugar, 3000 ppm SO2, 0.2% ascorbic acid and 1% citric acid; this method is described as producing the best dehydrated product. Drying is described using an electric cabinet through flow dryer operated at 60° C. The product showed no browning after 1 year of storage.

Drum-drying of mango purée is described as an efficient, economical process for producing dried mango powder and flakes. Its major drawback is that the severity of heat preprocessing can produce undesirable cooked flavours and aromas in the dried product. The drum-dried products are also extremely hydroscopic and the use of in-package desiccant is recommended during storage.

Canning. This preservation technology is described in various technological flow-sheets in this bulletin.

Mango bar or "fruit leather" is presented in various flow-sheets.

Guava Processing Technologies

Guava Purée

Guava purée is used in the manufacture of guava nectar, various juice drink blends and in the preparation of guava jam. The washed sound fruit is first passed through a chopper or slicer to break up the fruit and this material is fed into a pulper.

The pulper will remove the seeds and fibrous pieces of tissue and force the reminder of the product through a perforated stainless steel screen. The holes in the screen should be between 0.033 and 0.045 in. The machine should be fed at a constant rate to ensure efficient operation.

The puréed material coming from the pulper is next passed through a finisher. The finisher is equipped with a screen containing holes of approximately 0.020 in. The finisher will remove the stone cells from the fruit and provide the optimum consistency to the product.

Perhaps the best way to preserve the guava purée is by freezing and the material passing through the finisher can be packaged and frozen with no further treatment. It is not necessary to heat the product to inactivate enzymes or for other purposes. The material can be frozen in a number of types of cartons and cans; however, a fibre box with a plastic bag inside is commonly used and is probably the less expensive.

It is also possible to can and heat process the guava purée and this can be accomplished by heating the purée to 195° F in an open double bottom kettle, filling into cans, closing the cans, inverting the cans for a few seconds, followed by cooling. Cans should be cooled rapidly to approximately 100° F before they are cased and stacked into warehouses.

Guava Juice and Concentrate

Guava juice can be used in the manufacture of a clear guava jelly or in various drinks. A clear juice may be prepared from guava purée that is depectinised enzymatically. About 0.1% pectin-degrading enzyme is mixed into the purée at room temperature; heating of the product at approximately 120° F will greatly speed the action of the enzyme. After 1 hr. clear juice is separated from the red pulp by centrifuging or by pressing in a hydraulic juice press. A batch-type or continuous-flow centrifuge can be used on the depectinised purée with no further treatment.

The clear juice after centrifuge or after press (and subsequent filtration) can be preserved by freezing or by pasteurization in hermetically sealed cans.

For shipment to overseas markets it may be advantageous to concentrate either the purée or the juice.

Recent Trends in Fruit and Vegetable Processing

New products. The number and variety of fruit and vegetable products available to the consumer has increased substantially in recent years. The fruit and vegetable industry has undoubtedly benefited from the increased recognition and emphasis on the importance of these products in a healthy diet.

Traditional processing and preservation technologies such as heating, freezing and drying together with the more recent commercial introduction of chilling continue to provide the consumer with increased choice. This has been achieved by new heating (e.g. UHT, microwave, ohmic) and freezing (e.g. cryogenic) techniques combined with new packaging materials and technologies (e.g. aseptic, modified atmosphere packaging).

The overall trend in new fruit and vegetable products is "added value", thus providing increased convenience to the consumer by having much greater variety of ready prepared fruit and vegetable products. These may comprise complete meals or individual components. The suitability of products and packages for microwave re-heating has been an important factor with respect to added convenience.

Table: Numbers of new fruit and vegetable products

	1990	*1991*	*1992 Jan-June*
Frozen vegetable products	66	95	21
Chilled vegetable products	76	81	78
Heat processed vegetables	51	60	38
Heat processed fruits	13	14	5
Fruit juices and drinks	73	83	46
Potato crisps	32	33	16

Source: C. Dennis (1993)

New product development in the fruit and vegetable sector is most important in meeting the continued challenge of providing the consumer with choice and high quality products.

A Fresh Look at Dried Fruit

New fruit varieties and advance in drying technologies are putting a fresh twist on dried fruit applications. Fruits that have been introduced to the drying process include cranberries, blueberries, cherries, apples, raspberries and strawberries - not to mention the traditional mainstays of raisins, dates, apricots, peaches, prunes and figs.

Perceived as a "value-added" ingredient, dried fruit adds flavour, colour, texture and diversity with little alteration to an existing formula. The growing interest in ethnic cuisines in U.S.A. and the change to a more healthy way of eating, has also moved dried fruit considerably closer to the mainstream.

Found primarily in the baking industry, dried fruit is coming into its own in various food products, including entrees, side dishes and condiments. Compotes, chutneys, rice and grain dishes, stuffings, sauces, breads, muffins, cookies, deserts, cereals and snacks are all food categories encompassing dried fruit.

Since some dried fruit is sugar infused (osmotic drying), food processors can decrease the amount of sugar in formula - this is especially the case in baked products. Processors are making adjustments in moisture content of the dried fruit so that a varied range is available for different applications. An added bonus is dried fruits' shelf stability (a shelf life of at least 12 months). Dried fruit is more widely available in different forms, including whole dried, cut, diced and powders.

Citric acid and its use in fruit and vegetable processing. Citric acid may be considered as "Nature's acidulant". It is found in the tissues of almost all plants and animals, as well as many yeasts and moulds.

Commercially citric acid is manufactured under controlled fermentation conditions that produce citric acid as a metabolic intermediate from naturally-occurring yeasts, moulds and nutrients. The recovery process of citric acid is through crystallization from aqueous solutions.

Citric acid is widely used in carbonated and still beverages, to impart a fresh-fruit "tanginess". Citric acid provides uniform acidity, and its light fruity character blends well and enhances fruit juices, resulting in improved palatability. The amount of citric acid used depends on the particular desired flavour (e.g., High-acid: lemonade; Medium-acid: orange, punch, cherry; Low-acid: strawberry, black cherry, grape).

Sodium citrate is often added to beverages to mellow the tart taste of high acid concentrations. It provides a cool, distinctive smooth taste and masks any bitter aftertaste of artificial sweeteners. In addition, it serves as a buffer to stabilise the pH at the desired level. The high water solubility of citric acid (181 g/100 ml) makes it an ideal additive for fountain fruit syrups and beverages concentrates as a flavour enhancer and microbial growth inhibitor (preferably at pH < 4.6).

In processed fruits and vegetables, citric acid performs the following functions:

a. It reduces heat-processing requirements by lowering pH: inhibition of microbial growth is a function of pH and heat treatment. Higher heat exposure and lower pH result in greater inhibition. Thus the use of citric acid to bring pH below 4.6 can reduce the heating requirements. In canned vegetables, citric acid usage is greatest in tomatoes, onions and pimentos. For tomato packs, the National Canners Association recommends a pH of 4.1 to 4.3. In general, 0. 1% citric acid will reduce the pH of canned tomatoes by 0.2 pH units.

b. Optimise flavour: citric acid is added to canned fruits to provide for adequate tartness. Recommended usage level is generally less then 0.15%.

c. Supplement antioxidant potential: citric acid is used in conjunction with antioxidants such as ascorbic and erythorbic acids, to inhibit colour and flavour deterioration caused by metal-catalysed enzymatic oxidation. Recommended usage levels are generally 0.1% to 0.3% with the antioxidant at 100 to 200 ppm.

d. Inactivate undesirable enzymes: oxidative browning in most fruits and vegetables is catalysed by the naturally present polyphenol oxidase. The enzymatic activity is strongly dependent on pH.

Addition of citric acid to reduce pH below 3 will result in inactivation of this enzyme and prevention of browning reactions.

Cherry and Apricot Oils are Safe for Food Use

The oils obtained by cold pressing the kernels of the cherry (Prunus cerasus) and the apricot (Prunus armeniaca) have been declared acceptable for food use by the UK Ministry of Agriculture, Fisheries & Food's Advisory Committee on Novel Foods and Processes (ACNFP) by June 1993.

In its assessments of the safety in use of the cherry and apricot kernel oils, the ACNFP consider specifications that included data on fatty acid composition, the presence of natural antioxidants and the content of cyanide, mycotoxins and heavy metals.

The Committee says that it gave particular consideration to the possible presence in the oils of the cyanogenic glucoside amygdalin, from which cyanide is released by enzymic action when the kernels of cherry and apricot are crushed. Amygdalin was found to be absent from the cherry and apricot kernel oils.

The ACNFP is satisfied that there are no food safety reasons why the use of cherry and apricot kernel should not be acceptable provided there is compliance with the specifications shown in Table below.

Table: *Specification of purity for cherry and apricot kernel oils as determined by UK ACNFP*

	Cherry	*Apricot*
Contaminants limits: Heavy metals (total)	0.5 mg/kg	0.5 mg/kg
Aflatoxins (total)	4.0 g/kg	4.0 g/kg
Cyanide	0.15 mg/kg	0.15 mg/kg
Pesticide residues	0.01 mg/kg	0.01 mg/kg
Tocopherols:		
alpha/delta/gamma (mg/kg)	356-886	569-899

Source: Anon. (1993)

The oils are obtained by the mechanical mincing and cold pressing of kernels extracted from cleaned cherry or apricot stones. After filtering, the oils are stored and are to be sold in a raw, unrefined state. The cherry and apricot kernel oils are high unsaturated and are expected to be used as speciality oils for salad dressings, baking and shallow frying applications.

The Use of Fruit Juices in Confectionery Products

During the last decade, the concept of fruit juices has gained immensely on consumer popularity. The majority of new non-alcoholic and alcoholic fruit drink products were a combination of syrups, fruit juices and flavours.

The confectionery industry followed suit and new products incorporated fruit juices as part of their confectionery formulations and processes. Fruit juice concentrates of high solids are often used instead of normal or single-fold juices.

Juice concentrates are made of pure fruit juices. The process starts with pressing fruits and obtaining pure fruit juice; this is stabilised by heat treatment which inactivates enzymes and micro-organisms. The next processing step is concentration under vacuum up to 40-65° Brix or 4-7 fold. The concentrates are then blended for standardisation and stored.

These fruit juice concentrates are often further stabilised by the addition of sodium benzoate and potassium sorbate and are usually stored away from light and are refrigerated or frozen.

Depectinised fruit juices are also used to prevent foaming in confectionery processes and are essential for use in clear beverage products. Fruit juice concentrates which are depectinised, and have added preservatives are called stabilised, clarified, fruit juice concentrates.

Fruit juices are used in confectionery products in conjunction with natural and artificial flavours which provides intense flavour impact and are cost-effective for a confectionery product.

The traditional concern in using fruit juice concentrates in confectionery applications has been the effect of the natural acids on the finished product, particularly the formation of invert sugar during processing.

This is a logical concern since concentrates contain differing amounts and types of acids. For example: apple, cherry, strawberry and other berries contain primarily malic acid. Grapes mainly contain tartaric acid. Cranberry is high in quinic acid. Citrus fruits and pineapple contain differing amounts of citric acid. The concentrates, when used, are normally buffered to a pH of 5-7 with sodium hydroxide.

In formulating products with fruit juice concentrates, the solids of the concentrate are considered as mostly reducing sugars and a reduction in corn syrup is made to compensate for equivalent amount of reducing sugar being added in the concentrate.

The exact replacement can be determined by measuring the D.E. of the concentrate to be added. In formulations when small amounts of concentrate are used (less than 1%), no adjustment is made since the reducing sugar contribution of the concentrate is not significant.

Fruit juice concentrates can also be used to provide a source of natural colour, in particular red colour. Grape, raspberry, cherry, strawberry and cranberry concentrates in small amounts are very effective in colouring cream centres.

The inclusion of fruit juices in confectionery products is now left up to the imagination of the manufacturer. These products must, of course, hold up to the standards of flavour integrity, and product excellence, during the shelf-life of these products.

Chapter 9

Microbial Control by New Nonthermal Methods

Supermarket shelves today need to cater to the gourmet cook as well as the timeharried parent. Increasingly all types of consumers are demanding minimally-processed foods that are high in quality, nutritionally superior, and easy to prepare. Food processors have met this demand by developing refrigerated foods with extended shelf life. Ready-to-eat luncheon meats and complete heat-and-eat meals are some examples. By their very nature, however, these foods present challenges to ensure microbiological quality and safety.

This Scientific Status Summary addresses the microbiological concerns associated with extended shelf life refrigerated foods and control measures for ensuring microbiological quality and safety.

Extended shelf life refrigerated foods are foods that have received minimal processing or precooking and have an enhanced but limited shelf life; refrigeration is a key preservation measure. These foods include conventional products, such as luncheon meats and cured meats, as well as a new generation of partially processed refrigerated foods (NFPA, 1988) such as meat, seafood, egg, and vegetable salads, fresh pasta and pasta sauces, other sauces, soups, entrees, complete meals, and uncured meat and poultry items. Sous-vide foods, cooked inside a hermetically sealed plastic package under vacuum, are also included in this definition.

If extended shelf life refrigerated foods are heat processed, the heat treatment is much less than that required for commercial sterility. Canned foods are, therefore, excluded from this food category. Sous-vide foods and others that receive a lower heat treatment than that used for canning and that require refrigeration are described by some authors as "refrigerated processed foods of extended durability" (Peck, 1997).

Microbiological Concerns

The chief microbiological concerns associated with these products centre around two types of microorganisms— psychrotrophic and mesophilic pathogens—that could grow during extended refrigerated storage or temperature abuse. Psychrotrophs are bacteria, yeasts, and molds that grow, although slowly, at refrigeration temperatures (below 7°C) but grow optimally at temperatures above refrigeration, e.g., 25–30°C. Their maximum growth temperatures are 30–35°C (Kraft, 1992; Olson and Nottingham, 1980).

Mesophilic pathogens could survive under refrigeration and grow during any temperature abuse of the food. Mesophiles grow well between 20–45°C with optimum growth between 30–40°C (Jay, 1992). The potential for psychrotrophic spoilage microorganisms to grow during the extended refrigerated storage period and decrease organoleptic quality or spoil the food product is also a concern.

Pathogenic Microorganisms

Conventional wisdom of decades ago held that properly refrigerated foods would remain safe because it was thought that pathogenic bacteria could not grow at refrigeration temperatures. Microbial growth was thought necessary to either produce a sufficient number of cells or enough toxin to cause foodborne illness.

Since then, scientists have learned that several pathogens—such as Aeromonas hydrophila, nonproteolytic strains of Clostridium botulinum, Listeria spp., Yersinia enterocolitica, some strains of Bacillus cereus, enteropathogenic Escherichia coli, and Vibrio parahaemolyticus—can grow at refrigeration temperatures. Furthermore, scientists now know that some pathogens can cause illness when only a few cells are ingested. For example, as few as ten cells of the extremely virulent E. coli O157:H7 may cause hemorrhagic colitis (Buchanan and Doyle, 1997). Readers may refer to several texts for detailed information on the characteristics of pathogenic microorganisms and the foodborne illnesses they cause.

A. hydrophila is a facultative anaerobe that is generally considered a ubiquitous waterborne microorganism, occurring widely in fresh and brackish waters. Recent surveys (Pin et al., 1994; Saad et al., 1995; Schweizer et al., 1995) detected A. hydrophila in samples of raw milk, poultry, lamb, cheese, shellfish, pork, beef, watercress, lettuce, and escarole. Most cases of illness attributed to A. hydrophila have been sporadic, rather than associated with an outbreak (FDA, 1992).

C. botulinum is a ubiquitous anaerobic spore-forming bacterium whose spores are widely distributed in soil, freshwater and marine environments, raw agricultural products, and the intestinal tracts of fish and animals (Sugiyama, 1990). Four groups of C. botulinum (I-IV) and some strains of Clostridium baratii and Clostridium butyricum can produce botulinum neurotoxin (Hatheway, 1992). C. botulinum type I (proteolytic strains) and C. botulinum type II (nonproteolytic strains) are responsible for human foodborne botulism (Peck, 1997). The botulinum neurotoxins are differentiated as types A through G on the basis of serological reaction. The nonproteolytic strains—type E and some type B and F—do not produce overt signs of food spoilage during growth and toxin production.

Some nonproteolytic strains of C. botulinum are a concern with extended shelf life refrigerated foods because with sufficient time they may be able to grow and produce neurotoxin at temperatures as low as 3.3°C. Proteolytic strains, which are mesophilic, may be able to grow and produce toxin in foods if temperature abuse occurs. Most outbreaks of botulism in the United States have been caused by home-processed vegetables, fish, or meat products (ICMSF, 1996). The incidence of botulism from consumption of refrigerated foods is exceedingly low. However, the few outbreaks that have occurred and research challenge studies illustrate the potential C. botulinum hazards associated with extended shelf life refrigerated foods (Conner et al., 1989).

L. monocytogenes, a facultative anaerobe, is ubiquitous in the environment. L. monocytogenes has been isolated from soil, silage, food processing environments, and healthy humans and animals (ICMSF, 1996). A variety of foods, such as refrigerated ready to eat meat sandwiches and meat salads have been recalled from the marketplace because of contamination with L. monocytogenes (Ryser and Marth, 1991). Individuals with compromised immune systems, e.g., newborns, the elderly, and people suffering from the acquired immunodeficiency syndrome, are most susceptible to listeriosis. Outbreaks of listeriosis in North America have been associated with coleslaw, soft Mexican-style cheese, and milk (McLauchlin, 1996).

Y. enterocolitica is a facultative anaerobe whose main reservoir of bioserotypes pathogenic to humans is believed to be the pig (ICMSF, 1996). Symptoms of yersiniosis, the disease caused by Y. enterocolitica, may include fever, diarrhea, headache, vomiting, and severe abdominal pain similar to that associated with appendicitis. Y. enterocolitica has been isolated from a variety of animals, foods (lamb, pork, oysters, shrimp, and crabs), and water; however, isolates are often avirulent. Outbreaks of yersiniosis, which are relatively uncommon in the United States, have been caused by contaminated chocolate milk, recontaminated pasteurized milk, bean sprouts, tofu, and chitterlings (raw pork intestine).

B. cereus, an aerobic spore-former including psychrotrophic and mesophilic strains, is widely distributed in nature and in foods. B. cereus is commonly found in soil, milk, cereals, starches, herbs, spices, and other dried food products and on the surfaces of meats and poultry. B. cereus can produce two toxins that cause two distinct types of illness—a diarrheal illness and an emetic illness characterized by nausea and vomiting. Every well-documented report of B. cereus intoxication has described time and temperature abuse that enabled initially relatively low (innocuous) levels of B. cereus in foods to increase greatly. In most incidents, the food vehicle was a cereal or cereal- or spice-containing product (ICMSF, 1996).

A few serotypes and strains of E. coli, a facultative anaerobe that is part of the normal microflora of the intestinal tract of humans and most warm-blooded animals, can cause illness. Although they are not considered true psychrotrophs, some of these strains can grow at 6.9°C and below. Pathogenic E. coli are categorized into six groups—enteropathogenic, enteroinvasive, enterotoxigenic, enterohemorrhagic (EHEC), enteroaggregative, and diffusely adherent. Foods involved in outbreaks caused by pathogenic E. coli include meat, poultry, fish, vegetables, apple cider, raw milk, Brie and Camembert cheese, water, and radish and alfalfa sprouts. Some strains of E. coli, including some EHEC strains, are acid tolerant, a complex phenomenon that is growth phase dependent and inducible; acid tolerance may persist for extended periods at refrigeration.

V. parahaemolyticus is a facultatively anaerobic halophile (requiring sodium chloride for growth) occurring worldwide in inshore marine waters and frequently associated with molluscs, crustaceans, and fish (ICMSF, 1996).

Although the microorganism is considered mesophilic, growth has been demonstrated at temperatures as low as 5°C (Twedt, 1989). The microorganism is the most common cause of foodborne illness in Japan because it frequently contaminates seafood, often eaten raw in that country. Contaminated raw, improperly cooked, and cooked recontaminated fish and shellfish have been implicated in cases of gastroenteritis.

Spoilage Microorganisms

With sufficient time at refrigeration temperatures, several types of psychrotrophic bacteria, yeasts, and molds may grow to levels sufficient to cause food spoilage. These microorganisms include the Acinetobacter- Moraxella group, Alcaligenes species, Flavobacterium spp., Microbacterium spp., Xanthomonas spp., and the microorganisms of primary concern in extended shelf life refrigerated foods: Brochothrix thermosphacta, lactic acid bacteria (LAB), and Pseudomonas spp.

B. thermosphacta, which is aerobic (requiring free oxygen) to facultatively anaerobic (growing well either aerobically or anaerobically), has been recovered from vacuum-packaged beef, pork, lamb, and heat-processed cured meats such as sliced cooked ham, corned beef, and others (Kraft, 1992).

The extent of spoilage of vacuum-packed meat by B. thermosphacta varies with product pH and with oxygen permeability of the packaging material (Egan and Grau, 1981). Spoilage may involve development of sliminess and production of off-odors and off-flavours conferred by short chain fatty acids (Jay, 1992). The LAB, such as Lactobacillus spp., Leuconostoc, and Pediococcus, are facultatively anaerobic. The type of spoilage produced by the LAB is determined by the nature of the bacteria. Homofermentative LAB produce primarily lactic acid during sugar fermentation. Heterofermentative LAB produce acetic and formic acid, ethanol, and carbon dioxide, in addition to lactic acid. LAB can spoil a variety of foods, including milk and milk products, meats, vegetables, fruit juices, sugary products, alcoholic beverages, and products preserved with vinegar.

Pseudomonas spp., which are aerobic, are among the most common spoilage agents of refrigerated foods. Growth of Pseudomonas spp., like that of other Gram-negative psychrotrophs, is affected by oxygen tension, salt and other food additives, water activity (aw), pH, and other factors. During growth, pseudomonads produce proteases and lipases that can catalyze reactions causing degradation of protein and fat. The consequence of these reactions is formation of peptides and fatty acids of undesirable flavour (e.g., bitterness, rancidity) and odour. Sometimes these bacteria also produce unsightly green pigments.

Many yeasts and molds given sufficient time under refrigeration temperatures can spoil fruit juices, meat products, vegetables, dairy products, and possibly other foods. Some yeasts in the genera Candida,

Hanseniaspora, and Saccharomyces can grow in fruit juices at -5.5°C to -2.2°C, just above the freezing temperature for these foods (Pederson et al., 1959). Several genera of yeasts are found on fish and meat products. These include Candida, Cryptococcus, Debaromyces, Hansenula, Pichia, Rhodotorula, Saccharomyces, Sporobolomyces, Torula, Torulopsis, and Trichospora (Jay, 1987). Growth of yeasts on foods is commonly accompanied by production of carbon dioxide and yeasty, fruity, or alcoholic off-flavours and odors.

Psychrotrophic molds include Botrytis cinerea, Geotrichium candidum, Pullaria pullulans (Aureobasidium pullulans), and some species in the genera Alternaria, Monilia, Mucor, Penicillium, Sporotrichium, and Rhizopus. Not only does the visible presence of mold indicate spoilage, molds also commonly produce enzymes that degrade carbohydrates, fats, and proteins, causing softening of foods and flavour and aroma deterioration. Some species of molds, especially those in the genera Aspergillus, Fusarium, and Penicillium, can produce mycotoxins. Aspergillus spp. cannot produce mycotoxins at refrigeration temperatures, whereas certain species of Fusarium and Penicillium can (Frisvad and Samson, 1991).

Control Measures

Several types of control methods are effective in preventing or minimizing microbial contamination of product and inhibiting the growth of or destroying microbial contaminants.

Good Manufacturing Practices

(GMPs), Sanitation, and Hygiene. Processors need to select high-quality raw materials with low levels of microorganisms, especially psychrotrophs. They need to determine potential microbiological hazards of ingredients, possibly using microbiological specifications for ingredients to minimize risk (Moberg, 1989). Fabrication of raw materials into finished products under hygienic conditions is also important.

Food processing equipment must be designed and constructed so that it:

(1) is inert to the product,

(2) has smooth and nonporous product-contact surfaces,

(3) is readily accessible for cleaning and inspection,

(4) is self-emptying or self-draining,

(5) has covers to prevent external contamination, and

(6) has readily cleanable surfaces that do not contact the product and do not harbour contaminants (Cliver and Marth, 1990).

The equipment should be cleaned and sanitized as often as is necessary during a day's operation to prevent development of a biofilm that can contaminate subsequent lots of product. Cleaning and sanitizing adequacy can be determined using the more traditional swab procedures, the RODAC (agar contact) method, or the newer rapid ATP (adenosine triphosphate) bioluminescence assays.

Filtration of air entering the food processing area reduces the number of airborne contaminants. If processed foods will not receive a heat treatment or will have few barriers to microbial growth, use of "absolute" (high efficiency) air filters can virtually eliminate microbial contamination. If an air conditioning system is present, the system must be maintained properly so that condensate drains freely and does not contaminate the product. Food processing personnel must use hygienic practices and must be barred from moving from areas containing raw materials to areas containing finished products.

GMPs, sanitation, and hygiene are necessary prerequisites for implementing an effective Hazard Analysis Critical Control Point (HACCP) system, which enables the highest level of food safety assurance possible. HACCP is a systematic approach to the identification, evaluation, and control of food safety hazards, from raw material production and procurement to distribution and consumption of the finished product (NACMCF, 1997). HACCP is based on seven principles:

(1) conduct a hazard analysis,
(2) determine the critical control points,
(3) establish critical limits,
(4) establish monitoring procedures,
(5) establish corrective actions,
(6) establish verification procedures, and
(7) establish record keeping and documentation procedures.

Multiple Barriers/Hurdles

Referred to as the hurdle concept or hurdle technology, this approach combines several factors at subinhibitory concentrations that can effectively control microorganisms in refrigerated foods. Common hurdles include physical elements—such as refrigeration, modified atmosphere packaging (MAP), and heat treatment— and physicochemical factors—such as aw, pH, and preservatives. When used together, hurdles interact, sometimes synergistically, enabling use of lower intensities of each factor than would be necessary if each were used alone.

Scott (1989) recommended that challenge studies be conducted to verify the effectiveness of the combination of hurdles. For example, Simpson et al. (1995) demonstrated the antibotulinal effects of salt and pH in minimally processed sous vide spaghetti and meat sauce on proteolytic C. botulinum types A and B spores and toxin production.

The challenge studies indicated that the probability of toxigenesis increased with storage time, but that it decreased as either the aw or pH was decreased. Growth of the spores and toxin production were prevented in the product that was processed at 75°C for 36 min and held at 15°C (simulating mild temperature abuse) for 42 days by >1.5% salt (aw 0.983) and pH 5.5.

Ingredients

Some products can be formulated with ingredients that are barriers to microbial growth. For example, organic acids, particularly acetic but also lactic and citric, can inhibit bacterial growth. Sorbate, propionate, and benzoate have both antibacterial and antifungal properties. Although salt has antimicrobial properties it is not commonly used in high enough concentrations to be effective; this is particularly true of "low-sodium" foods. When appropriate, use of salt and other ingredients to reduce the aw to 0.980 or below will lengthen the lag phase of most bacteria and will further reduce the rate of any subsequent growth.

Heat Treatment

Heating foods will reduce the microbial population; the degree of reduction depends on the magnitude of the heat treatment, i.e., time and temperature. The magnitude of heat treatments commonly used is pasteurizing (destructive to vegetative pathogens) rather than sterilizing. Heat treatments lower than those producing commercial sterility are likely to inactivate vegetative cells, but not bacterial spores. Once the heating is completed, stringent hygienic measures must be implemented to prevent recontamination of the food with psychrotrophic spoilage or pathogenic microbes.

Modified Atmosphere Packaging

MAP extends product shelf life by reducing oxygen and/or increasing gases, such as carbon dioxide, in the food product environment. MAP inhibits the growth of aerobic spoilage microorganisms, such as Pseudomonas species, but allows facultative anaerobes such as LAB to grow.

Integrated with aseptic packaging, the technology has experienced rapid growth in the minimally processed refrigerated foods sector (Brody, 1996). Use of MAP is not without some risk, however. Any facultatively anaerobic or anaerobic psychrotrophic pathogens, such as nonproteolytic C. botulinum type E or Y. enterocolitica, may be able to grow until the LAB have reduced the pH of the product to inhibitory levels. Further, unlike aerobic spoilage microorganisms, growth of LAB may not be accompanied by overt evidence of spoilage. Moreover, if the MAP product received a nonsterilizing heat treatment, any surviving C. botulinum spores may, upon temperature abuse of the product, germinate, grow, and produce toxin without organoleptic signs of spoilage.

Storage Temperature and Shelf Life

Microbial lag phases (during which there is no growth or a decline in microbial numbers) and generation times (duration between formation of a daughter cell and its division into two new cells) increase as refrigeration temperature decreases. Thus, product temperature should be maintained just slightly above freezing. Acceptable product shelf life (e.g., days or weeks) at specified temperature limits should be established and monitored to help manage food quality and safety. Because the potential exists for temperature abuse at some point during handling or for storage past the intended shelf life, timetemperature indicators or integrators can be useful in determining when refrigeration temperatures or intended storage times have been exceeded.

Other Control Measures

Except for the limited use of food irradiation, the bacteriocin nisin, and high hydrostatic pressure (HHP), these methods are not yet fully developed nor commercially applied.

Ionizing radiation, from gamma rays (produced by the radioisotopes cobalt-60 or cesium-137), machine generated x-rays (with a minimum energy of 5 million electron volts, MeV), and electrons (with a maximum energy of 10 MeV) has been studied extensively. The United States has accepted this nonthermal processing technology for insect disinfestation of wheat, wheat flour, and fresh fruits and vegetables, inhibition of maturation of fresh fruits and vegetables, sprout inhibition of potatoes, inactivation of Trichinella spiralis in pork, and microbial decontamination of spices, herbs, vegetable seasonings, poultry, and red meats. Commercial application of ionizing radiation to foods in the United States, however, has grown slowly (Olson, 1998). Widespread application to refrigerated foods requires consumer acceptance.

Pulsed electric fields technology (PEF), also nonthermal, uses very short pulses of high intensity electric fields to inactive microorganisms. It has been largely limited to liquids such as juices, milk, and liquid egg. Applied to foods, PEF has the potential to equal conventional pasteurization (Yousef, 1996).

Pulsed high-intensity light is a nonthermal technology that uses a xenon flashlamp to generate extremely brief (< 2 msec) flashes of intense broad-spectrum (200 to 1,100 nanometer wavelength) light to inactivate microorganisms (yousef, 1996). accepted by the food and drug administration (fda, 1996) for controlling microorganisms on the surface of food, the technology is also useful for treating surfaces of equipment and packaging materials. successful commercial application to food requires further development because the method as currently understood suffers from limited penetration into food and may cause lipid oxidation (yousef, 1996).

HHP is effective in controlling microorganisms. Raffalli et al. (1994) demonstrated that L. innocua added to 35% fat cream at 107 cells per milliliter was reduced 98.7–99.99% after treatment at 25–26°C for 10–30 min. The D-value for L. innocua was 7.4 minutes. After treatment for 20 and 30 min, all surviving microorganisms were injured; a resuscitation step was needed before they were able to grow on a selective medium. HHP is applied commercially to refrigerated avocado products. Bacteriocins are antimicrobial proteins produced by certain bacteria.

The best known bacteriocin is nisin (also designated as a lantibiotic), produced by certain strains of Lactococcus lactis subsp. lactis (formerly Streptococcus lactis). In the United States, nisin is generally recognized as safe (GRAS) for limited use in pasteurized processed cheese to control growth of and toxin production by C. botulinum (FDA, 1988). GRAS petitions have been filed for use of nisin in reduced cholesterol liquid whole eggs, sauces, and nonstandardized salad dressings (FDA, 1994, 1995). In recent years, an array of other bacteriocins, many of which are inhibitory to foodborne pathogens, has been discovered.

These include unnamed bacteriocins produced by enterococci, pediocin produced by Pedicoccus acidilactici (Huang et al., 1994), bavaricin produced by Lactobacillus bavaricus, mesenterocin produced by Leuconostoc mesenteroides, carnocin produced by Carnobacterium piscicola (Bagi and Buchanan, 1994; Mathieu et al., 1994), sakacin produced by Lactobacillus sake (Holck et al., 1994), and curvaticin produced by Lactobacillus curvatus (Garver and Muriana, 1994).

Labeling

Manufacturers recognize the potential for temperature abuse during distribution or storage of foods requiring refrigeration. Hence, they voluntarily use label statements, such as "keep refrigerated" or "refrigerate after opening," to inform consumers of the need to maintain product at refrigeration temperatures (FDA, 1997).

The FDA determined in 1997 that the labeling of potentially hazardous foods that need refrigeration by consumers should be more specific about the types of hazards present and the necessary storage conditions after the food is opened and issued labeling guidance (FDA, 1997) to food manufacturers. The agency delineated foods that need refrigeration into three groups and developed model label statements and guidance on label placement and prominence. The model label statements refer to the importance of refrigeration for foods in two of the groups to maintain safety and the use of refrigeration for foods in the third group to maintain quality.

Forgive the pun, but there are some killer microbial intervention technologies on the market that are of keen interest to the food industry. The number of commercially available thermal, non-thermal, antimicrobial chemical and other novel treatments applied to foods and/or food-contact surfaces designed to inhibit, reduce or inactivate microorganisms has never been greater—or more in demand by food manufacturing operations.

What has driven a surge in the development and availability of new and improved interventions? Certainly, part of the answer is that regulatory, industry and research institutions have formed a strong partnership supporting the use of quantitative microbial risk assessments to identify critical points within the food chain at which effective interventions will have the greatest impact on reducing foodborne illness. Several of the U.S. Food and Drug Administration's (FDA) and the U.S. Department of Agriculture's (USDA) more recent industry guidelines, directives and proposed rules reflect this unified push to apply intervention technologies to improve food safety. FDA's 2004 Produce Safety Action Plan, for example, identifies as part of the agency's objectives the promotion of risk assessments to identify with industry the microbial interventions that will be most effective in produce operations.

Similarly, USDA's Food Safety and Inspection Service (FSIS) interim final rule on the control of *Listeria monocytogenes* in ready-to-eat (RTE) meat and poultry products (Directive 10,240.4) provides incentives for RTE producers to use post-lethality treatments, antimicrobials ingredients at formulation and other intervention technologies to significantly reduce the risk of the presence or growth of *Listeria* on these products.

And, industry has been vocal in expressing its desire that new intervention strategies receive expedited regulatory agency review, as companies search for viable options that will improve their ability to prevent, control and eliminate undesirable microorganisms through HACCP, GMP and other food safety programs.

As *Food Safety Magazine's* editors have traveled the country this year, we've been introduced to several innovative, new or improved technologies and systems designed for use as part of the multi-hurdle intervention approachto food safety. Here, we'll highlight some of these in-plant thermal, non-thermal, chemical and novel microbial intervention systems as a quick guide to the latest advances in this important area of food safety.

Hot Thermal Interventions

Applying heat to foods and food-contact surfaces, either as a validated kill step or as a pathogen reduction method, is the classic means by which to achieve processing, packaging and post-packaging intervention. From steam treatments to dry heat to cook steps, the higher the temperatures applied, the higher the confidence that bacteria have been destroyed. For some foods, however, heat treatments can have an adverse affect on the quality attributes of finished product, such as minimally processed fruits and vegetables. Some of the newer microbial intervention technologies are designed with this issue in mind and thus are capturing the attention of food manufacturers nationwide.

One such thermal technology, Biosteam Technologies, Inc.'s trademarked ThermoSafe system, is a pasteurization method that has proven effective in the area of surface decontamination of fruits and vegetables destined for further processing, including fresh juice, cut fruit and melons, fruit concentrates and smoothies. The company reports that research conducted by Food Safety Net Services, Ltd., an ISO/IEC 17025 accredited network of laboratories, shows that the

ThermoSafe process can be effective in reducing the microbial load of *Listeria monocytogenes, Salmonella* spp., *E. coli* O157:H7, and *Lactobacillus* spp. on the surface of fruits and vegetables by 5 logs, and in some cases, up to 9 logs. In addition to the safety factor of the process, studies indicate taste panels were able to determine a sweeter flavour from the steamed fruit over the control or unsteamed product.

In describing the process, Biosteam notes that the Thermo Safe process uses time and temperature settings, depending on the commodity, to facilitate the desired microbial intervention step. The system can serve to clean and sanitize both products and equipment in minutes, automatically ending the sanitizing cycle at a predetermined temperature. This technology is designed to provide the user with a constant supply of steam, whether in the form of a portable unit or a large scale in-line applications system.

Similarly, ALKAR-RapidPak's Surface Pasteurization Technology (RP-SP) is designed to produce safer food products without loss of desirable product sensory characteristics. This flash pasteurization system promises a 3-log reduction for Listeria monocytogenes, for example, with no formula change required and no effect on packaging films.

While a conventional pasteurization method essentially recooks product, taking between four to 20 minutes of application time that must be followed by re-chilling, the RP-SP is designed to achieve surface pasteurization in 1.5 seconds. The system module uses bursts of steam and a jostling mechanism to pasteurize all surfaces of food products, followed by the injection of rapid bursts of high-pressure steam.

"Cool" Interventions Emerge

Once considered "novel" technologies relegated to pilot plant research, non-thermal intervention systems are beginning to take their place in the real world of the food plant. A few of these technologies have been available to industry for awhile, but have recently experienced a resurgence of interest as application ranges have broadened and procedures for use are better understood by end-users. For example, since obtaining FDA GRAS status for use as a direct food contact antimicrobial agent in 2001, ozone is gaining many industry adherents as an additional intervention to strengthen food safety programs, such as HACCP and GMPs, in the processing plant and beyond. Ozone, which is an effective biocide against biofilms, bacteria, viruses, fungi and protozoa, has long been successfully used for no-rinse sanitation of food-contact surfaces and in clean-in-place (CIP) systems in food plants. Today, ozone also can be dissolved into water and used for direct contact on fruits and vegetables, raw and ready-to-eat meat and poultry, fish and commercial eggs, making this an intervention that can be used along the entire food production chain.

According to DEL Ozone, which manufactures the AGW Series of ozone-enriched water systems for sanitation and direct food contact applications, the primary benefit of ozone-enriched water is that it kills microbes as effectively as chlorine, sans residual chemicals. For example, DEL's ozone-enriched water efficacy testing, performed by the National Sanitation Foundation (NSF) under AOAC International methodology, shows that ozone dissolved in water can provide a 6-log reduction of *Staphyloccoccus aureus, Salmonella choleraesuis* and *Pseudomonas aeruginosa* in as little as 30 seconds (depending on the microorganism). The testing also shows that a 5-log reduction of *E. coli*, and a 4-log reduction of both *Listeria monocytogenes* and Campylobacter jejuni can be obtained. DEL Ozone systems enable operators to apply ozone in both aqueous and gaseous forms to fruits, vegetables, meat, poultry and fish products, and can be easily integrated into flume, cascade, or drench type operations, as well as into both ozone prewash and ozone wash sections. Ozone's effectiveness as an intervention goes beyond pathogens. In studies conducted by Silliker, Inc., Novazone patented ozone-based purification technologies have been shown to be effective in disinfecting, inhibiting or removing mold growth, reducing contamination and decay, and improving shelf life. In one study, Novazone applied gaseous ozone for one night in a 175,000-cubic-ft. cold storage room containing contaminated lemons with a measurable aerobic plate count of more than 3 million colony forming units (CFUs) of microorganisms per gram. The independent testing showed that the ozone treatment not only reduced contaminants by a 2-3 log count, but in the bin dump room, also reduced the yeast count from 900,000 CFUs (untreated) to 32,000 CFUs (treated) and the mold count from 400,000 CFUs (untreated) to 70,000 CFUs (treated).

Again, non-thermal intervention technologies promise the food industry effective pathogen reduction as well as enhanced product quality and shelf life. Ozone technology is no exception, according to The BOC Group, whose ozone-based pathogen control technology from its food safety technology and consulting division, Intervent, is receiving good response from clients such as the Norway-based Stolt Sea Farm, a $1 billion global aquaculture company that has installed the technology in its Los Angeles, CA and Stratford, CT plants.

The plants, which process sturgeon and salmon, have made the Intervent system a core component of their pathogen reduction strategies, and report that validation tests have confirmed that, in addition to reducing pathogens, ozone had a significant, positive impact on product quality and shelf life. As a result, Stolt plans to eventually introduce the technology in all of its processing plants. Continued research also has informed system design improvements to ensure the delivery of the correct amount of ozone reliably and safely to food products and food-contact surfaces to increase the intervention benefit. Many ozone technology developers, for example, now have in-depth material compatibility sheets to increase the ensure the good condition of equipment and components after treatment.

While aqueous ozone is compatible with many materials, including ceramic, glass, silicone and Teflon seals, and stainless steel piping, in the recent past, some end-users were unaware that aqueous ozone should not be applied on surfaces made with natural rubbers, polyurethane or resin-based plastics.

In an effort to provide informative materials on the technology's proper application, a new player in the market, Ozone Safe Food/ eFoodSafety.com, has launched an in-depth website to help industry enhance its understanding of this intervention. The company offers several equipment models that feature its patented Vortex Mixing System, which compresses the ozone into microscopic bubbles in the water solution. The result, says the company, is higher ozone residuals with less off-gas to achieve a faster, more effective application.

High pressure processing (HPP), an automated system that kills microorganisms by interrupting their cellular function without the use of heat, is another intervention that is becoming more widely available—and affordable for food processors. HPP emerged in the 1990s as a food pasteurization alternative to heat or irradiation processing but initial costs of the systems have prevented their widespread use. Studies show that the process also extends product shelf life by inactivating spoilage organisms, and when appropriately used, does not alter the texture, appearance or flavour of foods.

HPP is effective on food products such as RTE meat, cheese, guacamole, fruit juice, onions, salsa, oysters and other seafood. In addition, in preliminary studies reported in March 2005, FDA/USDA research collaborators found that one-minute HPP treatments of oysters significantly reduced hepatitis A virus from study samples. Researchers also indicated that the technology might be successfully used as an intervention against noroviruses in the future. As the first commercial tolling processor, American Pasteurization Co. (APC) has introduced what it describes as access to food processors with volumes too low to justify the investment in HPP equipment.

The company works with processors to incorporate HPP into their current distribution chain, so that food is not only treated at APC, but can be date-stamped, labelled, packaged and boxed on its way to distributors and retailers. APC aims to make HPP easier and more convenient by eliminating extra shipping steps for small- to mid-sized processing operations that are interested in supplementing their existing manufacturing practices and processes with an additional food safety intervention step. Avure Technologies Inc. recently introduced its next-generation 215L ULTRA HPP system for higher-volume plant operations, particularly for sliced RTE meat processors looking for a post-packaging lethality step that inactivates Listeria monocytogenes in accordance with the recent USDA FSIS directive. This latest design upgrades Avure's 215-liter pressure vessel with a dual-pump pressurization system that uses a new generation of pressure intensifiers to accelerate process steps. Along with an advanced PC-based control system that allows multiple units to be monitored and controlled from a central location, Avure also has shortened the system's cycle times to improve throughput.

Beginning in 1990, FDA approved irradiation (ionizing radiation) treatments as a safe and effective microbial reduction method for categories including spices, poultry and eggs, red meats, ground beef, seafood, sprouts, and fruits and vegetables. Currently, FDA is considering adding RTE meats, including hot dogs, deli ham and turkey and bologna, to the list of foods that can be treated with ionizing radiation. Earlier this year, USDA researchers with the Agricultural Research Service (ARS) Food Safety Intervention Technologies Research Unit reported that there may be other benefits to using this non-thermal method, which inactivates pathogens, parasites and spoilage organisms by destroying or damaging their cell membranes.

In studies on the use of irradiation to eliminate *Listeria monocytogenes* from RTE meats, the USDA researchers found that certain common food additives, such as a combination of acetic acid salts and lactic acid in bologna formulations, make *Listeria* more sensitive to radiation, thus allowing a significant reduction the dose of radiation needed to inactivate the pathogen to maintain product quality attributes.

The researchers noted that the combined treatments also prevented growth of spoilage microorganisms for two months in some RTE meats. "Ionizing radiation, when combined with common food additives, has the potential to significantly reduce the incidence of listeriosis associated with consumption of RTE meats in the U.S.," concluded the scientists.

Commercially available irradiation technologies, such as gamma processing technology pioneer MDS Nordion's Centurion and Sterigenics' electron beam, gamma and X-ray processing technologies are well established as effective microbial interventions, from receiving through processing to post-packaging applications. Given the findings of this new research, the food industry will likely see some enhancements of irradiation technologies and systems to incorporate additional things like food additives.

Chemical Interventions on the Rise

One of the most active areas of new technology development is chemical intervention systems, in the form of both direct food additives and secondary additives, or food contact-surface treatments. Mionix's Safe2O brand line of products makes the list, with both USDA approved, multi-platform pasteurization technologies and proprietary antimicrobial agent formulations. The Safe2O RTE 03 chemical intervention, for example, involves a unique blend of acidic calcium sulfate, lactic acid and propionic acid that kills bacteria on contact and prevents further growth of *Listeria monocytogenes* on various ham products and other RTE food products. The product is a validated one-step solution to achieve Alternative 1 in FSIS Directive 10,240.4.

Surface-friendly chemical solutions, such as peracetic or peroxyacetic acids, also are gaining popularity. These microbial interventions are compatible with a range of surface materials, providing antimicrobial efficacy as well as protection from corrosion or other chemical-caused damage to equipment. For example, AFCO's PeraSafe, composed of 15% peracetic acid and 10% H2O2, combines the proven efficacy of an oxidizer with an organic acid, reducing corrosion rates on machinery.

PeraSafe is used as a direct microbial intervention on meat and poultry carcasses and in fabrication applications, including ground beef, cuts and trims. Ecolab's EcoShield, its integrated interventions systems program that offers the food industry a holistic, multiple-intervention approach, combines antimicrobial chemical treatments, irradiation technology and sanitation control systems to reduce or eliminate pathogens. Among its chemical antimicrobials, Ecolab's Inspexx 100 and 200 peroxyacetic acid-based food surface treatments are recommended for use in poultry spray rinsing and chill water to reduce microbial contamination on processed poultry surfaces, and for use in carcass sprays to control microbes on red meat surfaces, respectively.

When used with certain dispensing equipment, notes the company, system performance is optimized and greater antimicrobial efficacy against Salmonella typhimurium, Campylobacter jejuni, Listeria monocytogenes and *E. coli* O157:H7 is achieved.

Ecolab also offers Tsunami, a peroxyacetic acid-based additive used for microbial control in fresh-cut, processed and post-harvest fruit and vegetable flume and wash water systems.

PURAC America's Purasal Opti.Form, a blend of sodium or potassium lactate and sodium diacetate, also is well-suited for the multi-hurdle intervention approach. Research by Oscar Mayer's R&D team has shown that the antimicrobial can effectively suppress the growth of Listeria monocytogenes in RTE meats.

More Interventions of Interest

Other novel or improved technologies and applications of interest to food processors provide the opportunity for intervention and reduction of overall bioburden as a way to further enhance food safety. For example, e-Merge Interactive's fluorescence-based real-time imaging technology, VerifEYE, which is a used by meat processors in the U.S. to instantly detect and verify trace amounts of organic contamination on meat surfaces that can harbour pathogens such as *E. coli* O157:H7 and Salmonella, provides a detailed display of surface contamination of incoming meat products, which can then be accurately and rapidly removed—before carcasses travel further into the processing plant.

The technology, which is available in a portable handheld version called the Solo, designed for spot-checking surfaces, as well as in an automated, in-line Carcass Inspection System (CIS), for automated inspection of entire carcasses, has been recognized as an effective tool for reducing contamination prior to the application of other microbial interventions during further processing.

Engelhard Corp. is building the antimicrobial agent directly into packaging film in an innovative intervention application. The company's Aseptrol antimicrobial, a chlorine dioxide formulation that rapidly kills bacteria, viruses and cysts, is being incorporated into antimicrobial films that will enhance long-term or residual control of microbial contamination, as well as potentially extend shelf life. According to Engelhard, applications for these antimicrobial films range from packaging meat, poultry and seafood, to fresh fruits, vegetables and foods in transit or storage from producers to processors or institutional users. Similarly, Next Conveyor (formerly Materials Systems Engineering) has built a preventive barrier to bacterial bioburden directly into its NEXGEN conveyors. The company incorporates antimicrobial belting and bacteria-killing components as part of its sanitary design of NEXGEN conveyors. By using Westlake Plastics Technologies' Silveron product, the patented silver-based antimicrobial technology belonging to Milliken Chemical, the next-generation conveyors are kept cleaner by preventing or reducing the formation of biofilm on the equipment surface. The silver ions provide the product with its broad antimicrobial characteristics and are contained in tiny ceramic crystals that release the silver as needed. The controlled release and dispersion of the additive throughout the components ensures long-term antimicrobial intervention.

AgION Technologies' AgION antimicrobial coating technology another silver ion delivery system being used in the manufacture of building and equipment components to inhibit microbial growth. Interesting applications include the use of AgION in hardware and plumbing used in the foodservice industry, and in commercial HVAC air handling units.

Julie Larson Bricher is Editorial Director of Food Safety Magazine.

KSU Process Protects Ready-to-Eat Beef from Pathogens

The good news for consumers is that there are more ready-to-eat (RTE) meat products on the market and that vacuum packaging has made it possible to keep them fresh in appearance and taste. The problem is that pathogenic bacteria can grow on the meat in this packaging at both room and refrigeration temperatures. The better news is that Food Safety Consortium researchers at Kansas State University have found that a solution of sodium citrate can inhibit the growth of the bacterium *Clostridium perfringens* on restructured roast beef.

RTE meats go through a mild heat treatment, but the treatment stimulates rather than reduces the growth of the bacterium in vacuum packaging. One way to prevent the problem would be to follow federal guidelines to sharply cool down the meat within five hours. But not all of the current refrigeration technology makes that possible.

So, according to KSU food science professor Dr. Daniel Y.C. Fung, "There is a need for additional secondary safety barriers in vacuum-packaged meat products that will help prevent the growth of anaerobic bacteria such as C. perfringens during cooling procedures."

C. perfringens is a common foodborne bacterium that the federal government reported was responsible for more than 6% of bacterial foodborne disease outbreaks in 2000. Its significance lies also in its designation by the Centers for Disease Control and Prevention as an organism that can produce toxin that could potentially be used in a bioterrorist attack.

Microbial Genomics: A New Tool to Increase Food Quality and Safety

There's a new discipline in town: genomics, a new field of science that analyzes and compares the complete genome (genetic material of an organism) of organisms or a large number of genes in a simultaneous fashion. When the media commented on the success of the human genome project, everybody expected genomics to greatly improve medicine.

Now genomics is also entering food production and processing. Microorganisms play important roles in our foods. Microbial Genomics can help us understand what microorganisms do and how they do it, in ways that were not previously possible, helping us to better understand how they can be manipulated for our benefit. Future foods will benefit from more efficient, less costly processing methods, better quality, enhanced freshness and longer shelf lives.

Consumer demand for fresh tasting products and convenience foods - such as ready to eat chilled multi-component products - is increasing rapidly. Therefore, processing methods for mild preservation and novel, non-thermal preservation technologies are increasingly needed.

However, a major limitation of classic food microbiology research is that the effect of preservation strategies on spoilage microorganisms can only be determined after a number of days and with separate determinations for each organism. This classic approach is very time-consuming since it consists in tasks such as colony counting of surviving microorganisms or toxin production on plates.

Hence, it is important to develop better knowledge and methods, enabling a quick and reliable prediction of the best processing conditions and a direct insight in the effectiveness of the conditions being applied.

Microbial Genomics in Food Production and Processing

Genomic technologies offer a new alternative to the classic approach. They allow quick identification of microorganisms present in the (raw) product and they could help to directly measure the total response of the target spoilage microorganisms to the applied preservation methods. By making use of this new approach, known as applied microbial genomics, it might be possible one day to reduce the number of experiments needed to measure all relevant responses. The tools used for that are small chips containing the information of thousands of genes of food spoilage micro-organisms, which are attached to a solid surface (like a glass slide) in a grid-like array.

These so called "microarrays" could enable in the long run the outcome prediction of a preservation treatment and the definition of additional preservation steps if necessary. As a result, process control could be improved significantly and the energy input into preservation processes could be reduced. This is hoped to result in improved sensory properties and significant energy savings (due to tailor-made process conditions) and decreased product losses.

An Example: Campylobacter Jejuni

A number of Campylobacter species, grouped under the name of thermophilic campylobacters, are "formidable" pathogens. Among them, Campylobacter jejuni is one of the world's most "successful" food poisoning bacteria and is probably responsible for more than twice as many cases of poisoning as Salmonella. Until recently, however, there has been relatively little research into why the bacterium is so virulent, but work in the UK is beginning to shed some light on the problem.

C. jejuni is thought to have 1,700 genes. Genomics is being used to explore the activity of individual genes and to look at the variety of proteins produced by the organism when it faces different environmental challenges.

The adaptability of C. jejuni derives from a powerful set of regulatory genes which enable it to change its metabolism rapidly depending on its environment, for example whether it is in contaminated raw chicken, or residing in the human gut. Applying microbial genomic research a food sample can easily be tested for the presence of those genes, which would indicate a contamination with C. jejuni. This test is significantly faster than conventional techniques, and might enable the use of sterilization methods less damaging to the food product.

Further Applications

Application of microbial genomics is not limited to preservation technologies. In principle, all processes in which living (micro-) organisms are involved are amenable to the concept.

Genomics of food microbes generates valuable knowledge that can be used for metabolic engineering, improving cell factories and development of novel preservation methods.

Furthermore, pre- and probiotic studies, characterization of stress responses, studies of microbial ecology and, last but not least, development of novel risk assessment procedures will be facilitated. Genomics technology can even be applied as measures for traceability from farm to table.

Chapter 10

Enzymes in Milk and Cheese Industries

We have been using enzymes for thousands of years to make various drinks and things to eat, such as cheese, yoghurt, bread, beer, and wine. Of course the people who first discovered how to make these knew nothing about enzymes. In this unit we can see how these amazing biological catalysts are used in making these traditional foodstuffs, and also in making new products.

In the field of biotechnology there are many industrial applications that result in biotech products that we use everyday at home. Some of these are food science applications that utilize enzymes to produce or make improvements in the quality of different foods. In the dairy industry, some enzymes are required for the production of cheeses, yogurt and other dairy products, while others are used in a more specialized fashion to improve texture or flavour. Five of the more common types of enzymes and their role in the dairy industry are described below.

Rennet

Milk contains proteins, specifically caseins, that maintain its liquid form. Proteases are enzymes that are added to milk during cheese production, to hydrolyze caseins, specifically kappa casein, which stabilizes micelle formation preventing coagulation. Rennet and rennin are general terms for any enzyme used to coagulate milk. Technically rennet is also the term for the lining of a calf's fourth stomach.

The most common enzyme isolated from rennet is chymosin. Chymosin can also be obtained from several other animal, microbial or vegetable sources, but indigenous microbial chymosin (from fungi or bacteria) is ineffective for making cheddar and other hard cheeses.

Limited supplies of calf rennet have prompted genetic engineering of microbial chymosin by cloning calf prochymosin genes into bacteria. Bioengineered chymosin may be involved in production of up to 70% of cheese products. While use of bioengineered enzymes spares the lives of calves, it presents ethics issues for those opposed to eating foods prepared with GEMs.

Other Proteases

Milk contains a number of different types of proteins, in addition to the caseins. Cow milk also contains whey proteins such as lactalbumin and lactoglobulin. The denaturing of these whey proteins, using proteases, results in a creamier yogurt product. Destruction of whey proteins is also essential for cheese production.

During production of soft cheeses, whey is separated from the milk after curdling, and may be sold as a nutrient supplement for body building, weight loss, and lowing blood pressure, among other things. There have even been reports of dietary whey for cancer therapies, and having a role in the induction of insulin production for those with Type 2 diabetes. Proteases are used to produce hydrolyzed whey protein, which is whey protein broken down into shorter polypeptide sequences. Hydrolyzed whey is less likely to cause allergic reactions and is used to prepare supplements for infant formulas and medical uses.

Lactase

Lactase is a glycoside hydrolase enzyme that cuts lactose into it's constituent sugars, galactose and glucose. Without sufficient production of lactase enzyme in the small intestine, humans become lactose intolerant, resulting in discomfort (cramps, gas and diarrhea) in the digestive tract upon ingestion of milk products. Lactase is used commercially to prepare lactose-free products, particularly milk, for such individuals. It is also used in preparation of ice cream, to make a creamier and sweeter-tasting product. Lactase is usually prepared from *Kluyveromyces* sp. of yeast and *Aspergillus* sp. of fungi.

Catalase

The enzyme Catalase has found limited use in one particular area of cheese production. Hydrogen peroxide is a potent oxidizer and toxic to cells. It is used instead of pasteurization, when making certain cheeses such as Swiss, in order to preserve natural milk enzymes that are beneficial to the end product and flavour development of the cheese. These enzymes would be destroyed by the high heat of pasteurization. However, residues of hydrogen peroxide in the milk will inhibit the bacterial cultures that are required for the actual cheese production, so all traces of it must be removed. Catalase enzymes are typically obtained from bovine livers or microbial sources, and are added to convert the hydrogen peroxide to water and molecular oxygen.

Lipases

Lipases are used to break down milk fats and give characteristic flavours to cheeses. Stronger flavoured cheeses, for example, the italian cheese, Romano, are prepared using lipases. The flavour comes from the free fatty acids produced when milk fats are hydrolyzed. Animal lipases are obtained from kid, calf and lamb, while microbial lipase is derived by fermentation with the fungal species *Mucor meihei.*

Although microbial lipases are available for cheese-making, they are less specific in what fats they hydrolyze, while the animal enzymes are more partial to short and medium-length fats. Hydrolysis of the shorter fats is preferred because it results in the desirable taste of many cheeses. Hydrolysis of the longer chain fatty acids can result in either soapiness, or no flavour at all.

Enzymes in Baking Industry

For decades, enzymes such as malt and fungal alpha-amylases have been used in bread-making. Rapid advances in biotechnology have made a number of exciting new enzymes available for the baking industry. The importance of enzymes is likely to increase as consumers demand more natural products free of chemical additives. For example, enzymes can be used to replace potassium bromate, a chemical additive that has been banned in a number of countries.

The dough for bread, rolls, buns and similar products consists of flour, water, yeast, salt and possibly other ingredients such as sugar and fat. Flour consists of gluten, starch, non-starch polysaccharides, lipids and trace amounts of minerals. As soon as the dough is made, the yeast starts to work on the fermentable sugars, transforming them into alcohol and carbon dioxide, which makes the dough rise.

The main component of wheat flour is starch. Amylases can degrade starch and produce small dextrins for the yeast to act upon. There is also a special type of amylase that modifies starch during baking to give a significant anti-staling effect

Table: *Lists some of the bread properties that can be improved using industrial enzymes.*

Enzyme	*Effect*
Amylase	Maximises the fermentation process to obtain an even crumb structure and a high loaf volume
Maltogenic alpha-amylases	Improves shelf-life of bread and cakes
Glucose oxidase	Oxidative reaction with gluten to make weak doughs stronger, drier and more elastic
Lipase	Modifies the natural lipids in flour to strengthen the dough
Lipoxygenase	Bleaching and strengthening dough
Xylanase	Dough conditioning. Easier dough handling and improved crumb structure
Protease	Weakens the gluten to give plastic properties required in doughs for biscuits
Asparginase	Reduces the amount of acrylamid formed during baking

Gluten is a combination of proteins that forms a large network during dough formation. This network holds the gas in during dough proofing and baking. The strength of this gluten network is therefore extremely important for the quality of all bread raised using yeast. Enzymes such as hemicellulases, xylanases, lipases and oxidases can directly or indirectly improve the strength of the gluten network and so improve the quality of the finished bread.

Flour Supplementation

Malt flour and malt extract can be used as enzyme supplements because malt is rich in alpha-amylases. Commercial malt preparations can differ widely in their enzyme activity, whereas an industrial enzyme is supplied with a standardised activity. The alpha-amylases degrade the damaged starch in wheat flour into small dextrins, which allows yeast to work continuously during dough fermentation, proofing and the early stage of baking. The result is improved bread volume and crumb texture. In addition, the small oligosaccharides and sugars such as glucose and maltose produced by these enzymes enhance the Maillard reactions responsible for the browning of the crust and the development of an attractive baked flavour.

Bread and cake staling is responsible for significant financial loss for both consumers and producers. For instance, every year in the USA bread worth more than USD 1 billion is discarded. However the main saving on prolonging the shelf life is actually savings in transportation and fuel costs due to a more efficient distribution.

Staling is associated with a loss of freshness in terms of increased crumb firmness, decreased crumb elasticity and loss of moistness.

Staling is believed to be due to changes in starch structure during storage. When the starch granules revert from a soluble to an insoluble form, they lose their flexibility; the crumb becomes hard and brittle. For decades, emulsifiers have been used as anti-staling agents. However, they actually have a limited anti-staling effect and are subject to special labelling rules.

By contrast, Novozymes' bacterial maltogenic alpha-amylases has been found to have a significant anti-staling effect. It modifies the starch during baking at the temperature when most of the starch starts to gelatinise. The resulting modified starch granules remain more flexible during storage. Bread produced with maltogenic alpha-amylase has a far softer and more elastic crumb than bread produced with distilled monoglycerides as emulsifiers.

Dough Conditioning

Flour contains 2.5-3.5% non-starch polysaccharides, which are large polymers (mainly pentosans) that play an important role in bread quality due to their water absorption capability and interactions with gluten. Although the true mechanism of hemicellulase, pentosanase or xylanase in bread-making has not been clearly demonstrated, it is well known that the addition of certain types of pentosanases or xylanases at the correct dosage can improve dough machinability yielding a more flexible, easier-to-handle dough. Consequently, the dough is more stable and gives better ovenspring during baking, resulting in a larger volume and improved crumb texture.

Normal wheat flour contains 1-1.5% lipids, both polar and non-polar. Some of these lipids, especially the polar lipids such as phosphorlipids and galactolipids are able to stabilise the air bubbles in the gluten matrix. The addition of functional lipases modifies the natural flour lipids so they become better at stabilizing the dough. This ensures a more stable dough in case of over-fermentation, a larger loaf volume, and significantly improved crumb structure. Because of the more uniform and smaller crumb cells, the crumb texture is silkier and the crumb colour appears to be whiter. It also reduces the need for addition of emulsifiers like DATEM and SSL that otherwise are commonly added to dough in order to stabilise it. This in turn means that emulsifiers can be removed from the label.

Chemical oxidants such as bromates, azodicarbonamide and ascorbic acid have been widely used to strengthen the gluten when

yield of juice from white grapes can be raised by 15%), subtle but generally beneficial changes in the flavour and, in the case of wine-making, shorter fermentation times.

The enzymes used in brewing are needed for saccharification of starch (bacterial and fungal a-amylases), breakdown of barley a-1.4- and a-1,3-linked glucan (b-glucanase) and hydrolysis of protein (neutral protease) t increase the (later) fermentation rate, particularly in the production of high-gravity beer, where extra protein is added. Cellulases are also occasionally used, particularly where wheat is used as adjunct to help break down the barley is b-glucans. Due to the extreme heat stability of the B. amyloliquefaciens a-amylase where this is used, the wort must be boiled for a much longer period (e.g. 30 minutes) to inactivate it prior to fermentation. Papain is used in the later post-fermentation stages of beer-making to prevent the occurrence of protein- and tannin- containing 'chill-haze' otherwise formed on cooling the beer.

Recently, 'light' beers, of lower calorific content, have become more popular. These require a higher degree of saccharification at lower starch concentrations to reduce the alcohol and total solid content of the beer. This may be achieved by the use of glucoamylase and/or fungal a-amylase during the fermentation. A new enzyme preparation of fungal pectin lyase (EC 4.2.2.10) was shown to be useful for the production of cranberry juice and clarification of apple juice in the food industry (Semenova et al., 2006). A comparative study showed that the preparation of pectin lyase is competitive with commercial pectinase products. The molecular weight of homogeneous pectin lyase was 38 kDa. Properties of the homogenous enzyme were studies. This enzyme was most efficient in removing highly esterified pectin.

The zygomycete microfungus R. microsporus var. microsporus produced a 1,3-1,4-beta-D-glucan 4-glucanhydrolase (EC 3.21.73) which was able to hydrolyse beta-D-glucan that contains both the 1,3-and 1,4-bonds (barley beta-glucans) (Celestino et al., 2006). Its molecular mass was 33.7 kDa.

Maximum activity was detected at pH values in the range of 4-5, and temperatures in the range of 50-60°C. The enzyme was able to reduce both the viscosity of the brewermash and the filtration time, indicating its potential value for the brewing industry Landbo et al, (2006) examined the clarification and haze-diminishing effects of alternative clarification strategies on black current juice including centrifugation and addition of acidic protease and pectinolytic enzyme preparation and gallic acid.

The extent of clarification and haze diminishment varied after individual treatment with five different acidic proteases, but one of the protease preparations, Enzeco, derived from Aspergillus niger, consistently tended to perform best. The individual and interactive effects on juice turbidity, total phenols, and total anthocyanin content of clarification treatment involving the use of two selected acid proteases (Enzeco and Novozyme 89L), a pectinase (Pectinex BE 3-L), and gallic acid were evaluated in a full factorial 2(4) experimental design. Haze development during cold storage decreases when gallic acid or any of the enzyme preparation was employed individually, but negative interaction effects resulted when the pectinase was employed in combination with any of the proteases.

After 28 storage days at 2°C, the lowest levels of haze formation were achieved when the Enzeco protease preparation, added at 0.025 g/L, was added with 0.050 g/L of gallic acid and allowed to react in the juice for 90 min. at 50°C. The corresponding anthocyanin reduction was approximately 12% (compared to approximately 30% with gelatin silica soltreatment). The data support the hypothesis that phenol-protein interactions are involved in juice turbidity development during cold storage of berry juices and demonstrate that pre-centrifugation and protease-assisted clarification show promise as an alternative, phenolic-retaining clarification strategy in black currant juice processing.

At monosodium glutamate (MSG) biosensor with immobilized L-glutamate oxidase (L-GLOD) has been developed and studied for analysis of MSG in sauces, soups, etc. (Basu et al., 2006). The immobilized enzymatic membrane was attached with oxygen electrode with a push-cap system. The detection limit of the sensor was 1 mg/dl and the standard curve was found to be linear up to 20 mg/dl. Response time of the sensor was 2 minutes. Cross-linking with glutaraldehyde in the presence of bovine serum albumin (BSA) as a spacer molecule has been used for immobilization. Optimization of the sensor was done with an increase in L-GOLD concentration (6.3-31.5 IU)(and also with increase in loading volume of enzyme solution (5-20 microL). Optimization of pH and temperature was also studied. The permeability of O_2 through different membrane was studied with and without immobolized L-GLOD. The enzymatic membrane was used for 20 measurements and stability of membrane was observed.

Enzymes in Meat Industries

Papain

Papain, also known as papaya proteinase I, is a cysteine protease (EC 3.4.22.2) enzyme present in papaya (*Carica papaya*) and mountain papaya (*Vasconcellea cundinamarcensis*).

Papain Family

Papain belongs to a family of related proteins with a wide variety of activities, including endopeptidases, aminopeptidases, dipeptidyl peptidases and enzymes with both exo- and endo-peptidase activity. Members of the papain family are widespread, found in baculovirus, eubacteria, yeast, and practically all protozoa, plants and mammals. The proteins are typically lysosomal or secreted, and proteolytic cleavage of the propeptide is required for enzyme activation, although bleomycin hydrolase is cytosolic in fungi and mammals. Papain-like cysteine proteinases are essentially synthesised as inactive proenzymes (zymogens) with N-terminal propeptide regions. The activation process of these enzymes includes the removal of propeptide regions. The propeptide regions serve a variety of functions in vivo and in vitro.

The pro-region is required for the proper folding of the newly synthesised enzyme, the inactivation of the peptidase domain and stabilisation of the enzyme against denaturing at neutral to alkaline pH conditions. Amino acid residues within the pro-region mediate their membrane association, and play a role in the transport of the proenzyme to lysosomes. Among the most notable features of propeptides is their ability to inhibit the activity of their cognate enzymes and that certain propeptides exhibit high selectivity for inhibition of the peptidases from which they originate.

Structure

The papain precursor protein contains 345aa, and consists of a signal sequence (1-18), a propeptide (19-133) and the mature peptide (134-345). The amino acid numbers are based on the mature peptide. The protein is stabilised by 3 disulfide bridges. Its 3D structure consists of 2 distinct structural domains with a cleft between them. This cleft contains the active site, which contains a catalytic diad that has been likened to the catalytic triad of chymotrypsin. The catalytic diad is made up of the amino acids - cysteine-25 (from which it gets its classification) and histidine-159. Aspartate-158 was thought to play a role analogous to the role of aspartate in the serine protease catalytic triad, but that has since then been disproved. None-the-less, it does appear to play a role.

Function

The mechanism by which it breaks peptide bonds involves deprotonation of Cys-25 by His-159. Asparagine-175 helps to orient the imidazole ring of His-159 to allow this deprotonation to take place.

Cys-25 then performs a nucleophilic attack on the carbonyl carbon of a peptide backbone. This frees the amino terminal of the peptide, and forms a covalent acyl-enzyme intermediate. The enzyme is then deacylated by a water molecule, and releases the carboxy terminal portion of the peptide. In immunology, papain is known to cleave the Fc (crystallisable) portion of immunoglobulins (antibodies) from the Fab (antigen-binding) portion.

Uses

Its utility is in breaking down tough meat fibers and has been utilized for thousands of years in its native South America. It is sold as a component in powdered meat tenderizer available in most supermarkets. Papain, in the form of a meat tenderizer such as Adolph's, made into a paste with water, is also a home remedy treatment for jellyfish, bee, and yellow jacket (wasps) stings; mosquito bites; and possibly stingray wounds, breaking down the protein toxins in the venom. Papain can be used to dissociate cells in the first step of cell culture preparations. A ten-minute treatment of small tissue pieces (less than 1 mm cubed) will allow papain to begin breaking down the extracellular matrix molecules holding the cells together. After ten minutes, the tissue should be treated with a protease inhibitor solution to stop the protease action (if left untreated, papain's activity will lead to complete lysis of the cells). The tissue must then be triturated (passed quickly up and down through a Pasteur pipette) in order to break up the pieces of tissue into a single cell suspension.

It is also used as an ingredient in various enzymatic debriding preparations, notably Accuzyme. These are used in the care of some chronic wounds to clean up dead tissue. Papain can also be found as an ingredient in some toothpastes or mints as teeth-whitener. Its whitening effect in toothpastes and mints is minimal, however, because the papain is present in low concentrations, and is quickly diluted by saliva. It would take several months of using the whitening product to have noticeably whiter teeth. It is the main ingredient of Papacarie, a gel used for chemomechanical dental caries removal. Besides the advantage of avoiding the use of rotary cutting tools, it does not interfere in the bond strength of restorative materials to dentin.

Immunoglobulins

An antibody digested by papain yields three fragments: two 50 kDa Fab fragments and one 50kDa Fc fragment. The papain-digested antibody is unable to promote agglutination, precipitation, opsonization, and lysis.

Bromelain

Bromelain can refer to one of two protease enzymes extracted from the plant family Bromeliaceae, or it can refer to a combination of those enzymes along with other compounds produced in an extract.

Extract Components

Bromelain extract is a mixture of protein-digesting enzymes—called proteolytic enzymes or proteases and several other substances in smaller quantities. The proteolytic enzymes are referred to as sulfhydryl proteases since a free sulfhydryl group of a cysteine side-chain is required for function. The two main enzymes are:

- Stem bromelain - EC 3.4.22.32
- Fruit bromelain - EC 3.4.22.33

The other substances typically include peroxidase, acid phosphatase, protease inhibitors, and calcium.

History

The first isolation of bromelain was recorded by the Venezuelan chemist Vicente Marcano (BU1 1.Phar. 5,77) in 1891 from the fruit of pineapple. In 1892, Chittenden, assisted by Joslin and Meara, investigated the matter fully (Trans. Conn. Acad. Arts Sci. 8, 281-308), and called it 'bromelin'. Later the term 'bromelain' was introduced and originally applied to any protease from any plant member of the plant family Bromeliaceae. Bromelain was first introduced as a therapeutic supplement in 1957. Research on bromelain apparently was first conducted in Hawaii but more recently has been conducted in countries in Asia, Europe and Latin America. Germany has recently taken a great interest in bromelain research; bromelain is currently the 13th most widely used herbal medicine in Germany.

Source

Bromelain is present in all parts of the pineapple plant (*Ananas sp.*), but the stem is the most common commercial source, presumably because it is readily available after the fruit has been harvested. Pineapples have had a long tradition as a medicinal plant among the natives of South and Central America.

Uses

Meat Tenderizing: Along with papain, bromelain is one of the most popular substances to use for meat tenderizing.

Today, about 90% of meat tenderizer is used in consumer households. Bromelain is sold in a powdered form, which is combined with a marinade, or directly sprinkled on the uncooked meat. The enzyme will penetrate the meat, and by a process called forking, cause the meat to be tender and palatable when cooked. If the enzyme is allowed to work for too long, the meat may become too "mushy" for many consumers' preferences. Cooked or canned pineapple does not have a tenderizing effect, as the enzymes are heat labile.

Some prepared meat products, such as meatballs, and commercially available marinades include pineapple and/or pineapple-derived ingredients. Apart from the mushiness of meat that has been over-tenderized, the activity of bromelain and similarly proteolytic plant enzymes can be undesirable where it is inappropriate. In dishes that depend on their protein content for important attributes, uncooked pineapple or its juice can be a nuisance. For example, some dishes such as brawn and jelly [called gelatin dessert (Jell-o) in the US] rely on the setting of gelatin.

They will not set if they contain raw pineapple or pineapple juice. Raw figs, papaya, and similarly proteolytic vegetable matter causes similar problems. To prevent the problem, the fruit should first be cooked, or at least parboiled sufficiently to inactivate the enzymes. Thorough heating to above about 65 degrees C will generally suffice. This problem does not affect dishes based on nonproteinaceaous gelling agents, such as agar, though such agents may be prevented from gelling by too much heating in acid recipes.

Actinidain

Actinidain (or actinidin) is a type of cysteine protease enzyme found in fruits including kiwifruit (genus Actinidia), pineapple, mango and papaya. This enzyme is part of the papain-like peptidase C1 family. It is an allergen and commercially useful as a meat tenderizer.

Amyloglucosidase

Also called glucoamylase, this enzyme acts on starch, dextrins and sugars by cleaving the a-1, 4 glycosidic linkages releasing stepwise from the end of the chain. It is widely used in the manufacture of glucose and for conversion of carbohydrates to fermentable sugars.

They play a major role in starch sugar & alcohol industries.

Glucoamylases: Fungal glucoamylase: is capable of hydrolyzing glucosidic linkages of starch. Glucoamylase and pullulanase blend: for higher glucose yields from starch

Cellulase

Cellulase acts on cellulose molecules by hydrolysing the b-1,4 glycosidic linkages. It largely produces cellobiose, which can ultimately yield glucose units, depending on the characteristic of the enzyme. Cellulases find wide application in the food & beverages industries.

Fungal cellulase:is primarily active on soluble forms of cellulase.

Catalase

Catalase is the enzyme that breaks down hydrogen peroxide to water and molecular oxygen. Catalase effectively removes the residual hydrogen peroxide, ensuring that the fabric is peroxide-free and mainly used in food industry and also in egg processing with other enzymes.

Glucanases

Glucanase act on b-1,3 and b-1,4 bonds in b-D-glucans. B-glucanases are of particular interest to the brewing industry, where they act on the glucans that impede clarification of wort and filtration of beer.

Beta-glucanase

Thermostable bacterial beta-glucanase:for the hydrolysis of cereal beta-glucan polysaccharides.It reduce hazing and can also affect foaming during storage.

Glucose Oxidase

Glucose and hexose oxidases catalyze reactions between glucose and oxygen, producing gluconolactone and hydrogen peroxide.And,they are mainly used in baking-food industries.

Hemicellulase

Hemicellulases act on hemicellulose (also called pentosan), a polymer of pentose sugars. They are mainly used in the baking (food & beverages) industry to improve the quality of dough, the softness of the crumb and volume. Fungal hemicellulase: for the hydrolysis of galactomannan gums and soluble cellulose.

Lipase

Lipase breaks down natural lipids and oils to free fatty acids and glycerol.In otherwise,Lipase is usefull in baking and egg processing with other enzymes.

Pancreatic Lipase: hydrolyzes insoluble fats and fatty acid esters to yield monoglycerides, glycerides, glycerol and free fatty acids.

Microbial lipases:for fat hydrolysis and flavour development.

Pectinase

This enzyme group has a heterogeneous collection of several activities in varying proportion. Pectinases act on pectins and their derivatives and play a major role in the food & beverages industry. In fruit drinks, pectinase is a unique enzyme for the peeling of citrus fruit. It can be used to produce fresh fruit segments from oranges and grapefruit and are less likely to damage fruits than conventional cutting. Segments look better and there is less production wastage. Pectinases cut pectin chains to lower viscosity, while glucose oxidase removes dissolved oxygen, decrasing oxidative browning

Protease

Proteases are enzymes that act on proteins and convert them to peptides and free amino acids. They play an important role in the food & beverages – baking industries. Depending on the application acid, neutral and alkaline proteases are available.

Rennet

Rennet, which is biologically prepared from the *Mucor* strain, is used as a milk coagulant in the preparation of cheese. The food & beverages industry, specifically the dairy industry, has accepted microbial rennet as the next best thing to natural rennet.

Tannase

Tannase is an esterase-based formulation that hydrolyses the acyl esters of tannins, making tannins more soluble at a lower temperature and pH. They are used in the production of instant tea (for solubility) in the food & beverages industry.

Xylanase

Xylanases cleave chains of b-1,4-xylosidic linkages in Xylans. They are mainly used in the food & beverages- baking industries.

- Intramolecular oxidoreductases:
  - Interconverting aldoses and ketoses
  - Interconverting keto-and enol-groups
  - Transferring other groups.
- Intramolecular transferases:
  - Transferring acyl groups
  - Transferring phosphoryl groups
  - Transferring other groups.
- Intramolecular lyases:
  - Other isomerases.

6. Ligases:
   - Forming C-O bonds:
     - Amino-acid-RNA ligases.
   - Forming C-S bonds:
     - Acid-thiol ligases.
   - Forming C-N bonds:
     - Acid-ammonia ligases (amide synthetases)
     - Acid-amino-acid ligases (peptide synthetases)
     - Cyclo-ligases
     - Other C-N ligases
     - C-N ligases with glutamine as N-donor.
   - Forming C-C bonds.

Partial List of Enzyme Preparations that are Used in Food Industry

Food ingredients may be "food additives" that are approved for specific uses or GRAS (generally recognized as safe) substances. A substance may be GRAS only if its general recognition of safety is based on the views of experts qualified to evaluate the safety of the substance. GRAS status may be based either on a history of safe use in food prior to 1958 or on scientific procedures, which require the same quantity and quality of evidence as would be required to obtain a food additive regulation.

The following list,includes approved food additives and substances.This list includes some ingredients which have been the subject of opinion letters to individuals who asked whether object to the use of the ingredient in food on the basis of an independent GRAS determination. In addition, enzyme preparations may be the subject of a GRAS notice.

The following is a compilation of enzymes listed as food additives and is a compilation of enzymes that have been specified or un specified food uses and predicated on the use of nontoxicogenic strains of the respective organism and on the use of current good manufacturing practise.Please be awera that all food ingredients has not been considered.Therefore,this does not represent a complete list of all enzymes for some uses.

Amyloglucosidase derived from *Rhizopus niveus* for use in degrading gelatinized starch into constituent sugars and permitted in or upon distellers'mash;mash destined for vinegar manufacture and brewers'mash.

Carbohydrase and cellulase derived from *Aspergillus niger* for use in clam and shrimp processing.

Carbohydrase derived from *Rhizopus oryzae* for use in the production of dextrose from starch.

Catalase derived from *Micrococcus lysodeikticus* for use in the manufacture of cheese.

Esterase-lipase derived from *Mucor miehei* var.as a flavour enhancer in cheeses, fats and oils, and milk products.

Galactosidase derived from *Mortierella vinaceae* var. *raffinoseutilizer* for use in the production of sucrose from sugar beets.

Milk-clotting enzymes, microbial for use in the production of cheese (Milk-clotting enzymes are derived from *Endothia parasitica Bacillus cereus*, *Mucor pusillus lindt* and *Mucor miehei* and *Aspergillus oryzae* modified to contain the gene for aspartic proteinase from *Rhizomucor miehei* var) and permitted in or upon cheese, cottage cheese, sour cream and Emmentaler cheese.

Flour may contain -amylase obtained from the fungus *Aspergillus oryzae*.

Alpha-amylase enzyme preparation from *Bacillus stearothermophilus* used to hydrolyze edible starch to produce maltodextrin and nutritive carbohydrate sweeteners.

Bromelain derived from pineapples, *Ananas comosus* and *Ananas bracteatus* used to hydrolyze proteins and polypeptides; and permitted in or upon beer, bread, edible collagen sausage casing, milk, meat cuts and tenderizing preparations.

Mixed carbohydrase and protease enzyme product derived from *Bacillus licheniformis* for use in hydrolyzing proteins and carbohydrates in the preparation of alcoholic beverages, candy, nutritive sweeteners and protein hydrolysates. Catalase from bovine liver used to decompose hydrogen peroxide and permitted in or upon soft drinks and egg albumen. Ficin (peptide hydrolase) from the genus *Ficus* to hydrolyze proteins and polypeptides, and permitted in or upon beer, malt liquor, edible sausage casing, hydrolysed milk protein, meat cuts and meat tenderizing preparations.

Insoluble glucose isomerase enzyme preparations are derived from recognized species of precisely classified, nonpathogenic, and nontoxicogenic microorganisms, including *Streptomyces rubiginosus*, *Actinoplane missouriensis*, *Streptomyces olivaceus*, *Streptomyces olivochromogenes* and *Bacillus coagulans* grown in a pure culture fermentation that produces no antibiotic and permitted in or upon glucose to be partially or completely isomerized to fructose.

Lactase enzyme preparation from *Candida pseudotropicalis* for use in hydrolyzing lactose to glucose and galactose.

Lactase enzyme preparation from *Kluyveromyces lactis* (previously called *Saccharomyces lactis*) for use in hydrolyzing lactose in milk and permitted in or upon lactose-reducing enzyme preparations and milk destined for use in ice cream mix.

Animal lipase (triacylglycerol hydrolase) derived from the edible forestomach of calves, kids or lambs used to hydrolyze fatty acid glycerides and permitted in or upon dairy based flavouring preparations, liquid-dried egg white and Romano cheese.

Lipase enzyme preparation from *Rhizopus niveus* used in the interesterification of fats and oils.

Malt (-amylase and -amylase) from barley to hydrolyze starch.

Pancreatin (peptide hydrolase) from porcine or bovine pancreatic tissue used to hydrolyze proteins or polypeptides and permitted in or upon liquid–dried egg white, precooked cereals, starch used in the production of dextrins, maltose, dextrose, and glucose.

Papain derived from papaya, *Carica papaya* and permitted in or upon beer, malt liquor, beef before slaughter, edible callogen sausage casing, hydrolised animal, milk and vegetable protein, meat cuts and meat tenderizing preparations and precooked cereals.

Pepsin (peptide hydrolase) from hog stomach used to hydrolyze proteins and permitted in or upon beer, malt liquor, cheese, cottage cheese, cream cheese, defatted soya flour and precooked cereals.

Rennet (animal derived) and chymosin preparation from *Escherichia coli* K-12, *Kluyveromyces marxianus* var. *lactis* or *Aspergillus niger* var. *awamori* to coagulate milk in cheeses and other dairy products.

Trypsin (peptide hydrolase) from porcine or bovine pancreas used to hydrolyze proteins. Urease enzyme preparation from *Lactobacillus fermentum* for use in the production of wine. Aminopeptidase enzyme preparation from *Lactococcus lactis*used as an optional ingredient for flavour development in the manufacture of cheddar cheese.

Amiloglucosidase; malt, from *Aspergillus niger var.; Aspergillus oryzae var.; Rhizopus oryzae var.;* permitted in or upon beer, malt liqour; bread, flour; chocolate syrups; istellers' mash; precooked cereals; starch used in the production of dextrins, altose, extrose, glucose; unstandardized bakery products. Glucanase enzyme from *Aspergillus niger var.;Bacillus subtilis var.* permitted in or upon beer, malt liquor; corn for degermination; distellers' mash; mash destined for vinegar manufacture and unstandardized bakery products. Glucose oxidase enzyme from *Aspergillus niger var.* permitted in or upon soft drinks; liquid whole egg, egg white and liquid egg yolk destined for drying.

Hemicellulase enzyme from *Bacillus subtilis var.* permitted in or upon distellers'mash; liquid coffee concentrate and mash destined for vinegar manufacture. Invertase enzyme from *Saccharomyces sp.* permitted in or upon soft-centered and liquid-centered confections;and unstandardized bakery foods. Lipoxidase enzyme from soyabean whey or meal permitted in or upon beread;flour.

Pectinase enzyme from *Aspergillus niger var.; Rhizopus oryzae var.* permitted in or upon cider, wine; distellers'mash; fruit juice; natural flavour and colour extractives; skins of citrus fruits destined for jam, marmalade and candied fruit production and vegetable stock for use in soups. Pentosanase enzyme from *Aspergillus niger var.; Bacillus subtilis var.* permitted in or upon beer, malt liquor; corn for degermination; distellers' mash; mash destined for vinegar manufacture and unstandardized bakery products. Rennet enzyme from fourts stomach of calves, kids, or lambs; permitted in or upon cheese, cottage cheese, cream cheese; and cheese soread.

The following is a compilation of microbially derived enzymes. The opinions are predicated on the use of nonpathogenic and nontoxicogenic strains of the respective organisms and on the use of current good manufacturing practice.

Carbohydrase, cellulase, glucose oxidase-catalase, pectinase, and lipase from *Aspergillus niger.*

Carbohydrase and protease from *Aspergillus oryzae.*

Carbohydrase and protease from *Bacillus subtilis.*

Invertase from edible baker's yeast or brewer's yeast (*Saccharomyces cerevisiae*).

Enzyme Classification According to their Sources

Biologically active enzymes may be extracted from any living organism. A very wide range of sources are used for commercial enzyme production from *Actinoplanes* to *Zymomonas*, from spinach to snake venom. Of the hundred or so enzymes being used industrially, over a half are from fungi and yeast and over a third are from bacteria with the remainder divided between animal (8%) and plant (4%) sources. A very much larger number of enzymes find use in chemical analysis and clinical diagnosis. Non-microbial sources provide a larger proportion of these, at the present time. Microbes are preferred to plants and animals as sources of enzymes because:

1. they are generally cheaper to produce.
2. their enzyme contents are more predictable and controllable,
3. reliable supplies of raw material of constant composition are more easily arranged, and
4. plant and animal tissues contain more potentially harmful materials than microbes, including phenolic compounds (from plants), endogenous enzyme inhibitors and proteases.

Attempts are being made to overcome some of these difficulties by the use of animal and plant cell culture.

- The names in common usage are given. As most industrial enzymes consist of mixtures of enzymes, these names may vary from the recommended names of their principal component. Where appropriate, the recommended names of this principal component is given below.
- The EC number of the principal component.
- I - intracellular enzyme; E - extracellular enzyme.
- +++ > 100 ton year^{-1}; ++ > 10 ton year^{-1}; + > 1 ton year^{-1}; - < 1 ton year^{-1}.
- triacylglycerol lipase;
- chymosin;
- Endo-1,3(4)-b-glucanase;
- xylose isomerase;

- subtilisin;
- a-dextrin endo-1,6-a-glucosidase;
- glucan 1,4-a-glucosidase;
- b-galactosidase;
- microbial aspartic proteinase;
- polygalacturonase;
- a-galactosidase;
- b-fructofuranosidase.

In practice, the great majority of microbial enzymes come from a very limited number of genera, of which *Aspergillus* species, *Bacillus* species and *Kluyveromyces* (also called *Saccharomyces*) species predominate. Most of the strains used have either been employed by the food industry for many years or have been derived from such strains by mutation and selection. There are very few examples of the industrial use of enzymes having been developed for one task. Shining examples of such developments are the production of high fructose syrup using glucose isomerase and the use of pullulanase in starch hydrolysis.

Producers of industrial enzymes and their customers will share the common aims of economy, effectiveness and safety. They will wish to have high-yielding strains of microbes which make the enzyme constitutively and secrete it into their growth medium (extracellular enzymes). If the enzyme is not produced constitutively, induction must be rapid and inexpensive. Producers will aim to use strains of microbe that are known to be generally safe. Users will pay little regard to the way in which the enzyme is produced but will insist on having preparations that have a known activity and keep that activity for extended periods, stored at room temperature or with routine refrigeration. They will pay little attention to the purity of the enzyme preparation provided that it does not contain materials (enzymes or not) that interfere with their process. Both producers and users will wish to have the enzymes in forms that present minimal hazard to those handling them or consuming their product. The development of commercial enzymes is a specialised business which is usually undertaken by a handful of companies which have high skills in

1. screening for new and improved enzymes,
2. fermentation for enzyme production,
3. large scale enzyme purifications,
4. formulation of enzymes for sale,

5. customer liaison, and
6. dealing with the regulatory authorities.

Use of Enzymes in the more Comman Food Industrial Areas

Alcohol	Amylases, Amyloglucosidase, β-glucanase, Cellulase, Cellobiase, Pectinase and Protease
Amino Acids	Protease, Peptidases and Amidases
Biscuits	Amylases and Proteases
Bread and flour	Amylases, Amyloglucosidase, Cellulase, Glucanase, Lipase and Pentosanase, Xylanase, Glucose oxidase
Brewing	Amylases, Amyloglucosidase, Cellulase, Glucanase, Protease and Pentosanase
Butter	Glucose oxidase and Catalase
Celulose	*for glucose production;* Cellulase and cellobiase *for extraction;* Pectinase, Amylase and lignase
Coffee	Pectinase, Hemicellulase and cellulase
Cheese	*for flavour*; Lipase, Peptidase, Protease *for production;* Rennets *for ripening*; Protease, Peptidase, Lipase
Chocolate	*for soft centered chocolates;* Milk protein-based aerating
Confectionery	*for candies and softness;* Invertase *for gum products;* Amylase for *reworking;* Pectinase, Amylase, Protease, Invertase
Dairy products	*for lactose conversion*; Lactase for *protein modification*; Protease for *heat treated milk*; Sulphydryloxidase
Effluent treatment	Amylase, Protease, Cellulase, Pectinase, Lipase
Egg processing	*for liquid;* Protease for *whippining and emulsifying;* Lipase, Phospholipase for *preservation of albumen;* Glucose oxidase/catalase
Fats	Lipase, Esterase, Oxidase
Fish	*for cleaning curing and waste utilization;* Protease
Flavours	*for yeasts and proteins; Protease*, Peptidase, Glucanase *for carbohydrates;* Amylase, Protease *for fats;* Lipases, Esterases
Fructose	*from starch;* Glucose isomerase from *inulin;* Inulinase
Fruit,candied	Cellulase, Amylase, Pectinase, Protease
Fruit clouds	*for production;* Pectinase, Amylase *for removal;* Amylase, Pectinase, Cellulase

Contd...

Fruit colour control	Anthocyanase
Gluconic acid	Glucose oxidase/catalase
Glucose	*for production;* see"starch *for isomeration;* see"fructose
Ice cream	Lactase, Lipase
Meat tenderization	Protease, Peptidase
Nucleic acid	*for removal;* Nuclease *for flavour derivatives; Phospho*-esterase
Oils	*for animal oils;* Protease, Esterase *for Mineral oils;* Protease, Cellulase, Amylase, Pectinase, Hemicellulase
Protein	*for extraction;* Amylase, Cellulase, Hemicellulase, Glucanase, Pectinase *for modification;* Protease, Peptidase, Esterase
Starch	Amylase (for *saccharification to maltose*), Amyloglucosidase, Cellulase, Hemicellulase, Glucanase, Pullulanase, Lipase, Pectinase, Protease, Isomerase (for *isomerization*), Glucoamylase (*for saccharification to glucose*)
Sucrose	*for production;* Dextranase, Amylase, Galactosidase *for inversion;* Invertase
Tea	*for fermentation;* Cellulase, Glucanase, Pectinase *for solubility;* Tannase
Whey	see" dairy and protein
Wine	*for production;* Amylase, Cellulase, Pectinase *for clarification;* Amylase, Glucanase, Pectinase *for colour control;* Glucose oxidase/catalase
Extraction Filtration (improvement)	Cellulase
Milard reaction (to prevent)	Glucose oxidase
Grain Processing	Glucoamylase, Amylase, Glucose isomerase,

Enzimas (GMOs)

Industrial enzymes are used in a wide range of foods, to improve gains, processability, shelf-life, taste and other properties in flour, starches, pop, drinks, fruit juices, oils, beer, whine, cheese and meat. These artificially-added enzymes don't have to be declared on the labels, and it is hard to avoid them. Many of these enzymes are produced by genetically modified organisms (GMOs), usually molds and bacteria. Since the produced enzymes are subsequently separated from the GMOs, the use of GMOs doesn't have to be declared. However, the separation is often incomplete, and residuals of the molds and bacteria are the main culprits of allergies.

New Enzymes, their Efficiencies and the Keys

Plasmin Protease Enzyme

Plasmin protease enzymes present in whey ingredients may lead to improved texture in yogurt or better flavour and texture in natural cheese.The enzymes, which originate in bovine blood, may also have negative effects, such as decreasing the water binding capacity of processed meats and producing undesirable textures in pasteurized processed cheese. The research showed that commercial whey protein isolates generated by ion-exchange separation tend to have the lowest plasmin levels. Protein concentrates from sweet whey (rennet coagulated curd) have intermediate amounts, and the highest levels occur in concentrates from acid whey (produced by acid precipitated cheese curds).

Although bacterial growth during raw milk storage decreases the plasmin content in the casein portion (i.e., cheese curd) and increases the content in whey, whey separation methods have an even more profound effect.

Ferulic Acid Esterase

Ferulic acid esterase (FAE) is a recently discovered enzyme with a very broad application range. Ferulic acid esters are formed in cell walls as ester linkages between hydroxy substituents of the xylan (carbohydrate) chain and acid residues from the ferulic acids. Additionally, the ferulic acid moieties may photodimerise, which further crosslinks the xylan chains and results in the carbohydrate components becoming inaccessible to enzymatic hydrolysis. In general, ferulic acid esterase is involved in hydrolysing recalcitrant parts of the plant cell wall, the beginning of a process that enables more complete breakdown of fibrous or pectinaceous cell walls and consequently the release of a higher proportion of sugars or cell wall bound flavours. The increase in the hydrolysis of the cell wall is brought about because the ferulic acid esterase breaks bonds that restrict accessibility of other cell wall hydrolases to their substrate. Consequently ferulic acid esterase gives improvements in any process where a more complete breakdown of the cell wall, or increase in the release of free sugars is an advantage.

The specific bond that is hydrolysed is the ester bond between a phenolic acid moiety and a sugar, which is linked to the xylose backbone of the cell wall, usually an arabinose residue (and sometimes a galactose residue). Hydrolysis, therefore, results in the production of a free phenolic acid, which may be used in the production of other commercially significant compounds such as p-coumaric acid, which is used as a photoactive ingredient in sunscreens, or a further biotransformation which will give rise to natural vanillin.

Fruit juice and vegetable processing also benefit from the inclusion of that. The resulting increased cell wall breakdown means that the filtration of the final product is much easier due to the reduction in particulate matter present. Ferulic acid esterase would benefit in carrot processing, whether the desired result is the complete liquifaction of the carrot, or the extraction of juice.

Flavour extraction may also be enhanced by Ferulic acid esterase because of the more complete hydrolysis of the cell wall that is attained. This enables more of the entrapped flavour components to be released in many extraction processes. Although processes vary considerably, this enzyme is active in the temperature range of 30 - 50¾C and at mildly acidic to neutral pH. The other modified type of Ferulic acid esterase can benefits when used in brewing: more fermentable sugars can be released which means that the amount of substrate for the fermentation can be reduced. Waste processing of natural products, e.g.. corn, wheat, sugar beet can be carried out using this enzyme, in conjunction with other cell wall hydrolases. In addition to the advantages of having less solid material to transport at the end of any processing that has the ability to release free phenolic acids from the waste material, which with a simple biotransformation can give rise to value added products such as p-coumaric acid a common protectant used in sun cream and vanillin.

Transglutaminase (TG)

Firstly, transglutaminase enzyme approved for use in all non-standardized processed meat and poultry products. And after, it approved its use in processed seafood. The enzyme, TG, crosslinks protein molecules in food to form firmer textured meat and seafood products without using binders or salt and for use in meat substitutes as well as refrigerated and frozen yogurt products and also nucleotides synergize with MSG to deliver "umami," the fifth basic taste.

The approved enzyme for use in meats, poultry, seafood and cheese, a unique transglutaminase (TG) enzyme gives food formulators the opportunity to be creative. For example, when caseinate is mixed with water, it will not gel. However, with it added the transglutaminase crosslinks the protein, causing it to gel without heating. TG also has the potential to join two different proteins that normally are not linked together. It gives one the potential to create designer proteins.

Transglutaminase enzymes from different sources such as animal tend to require calcium in order to be active. Being calcium independent makes it easier to use TG in a wide range of products. In addition to its ability to create new products, TG used for restructuring products, which can often lead to cost savings. In the U.S., Ajinomoto sells TG in preparations for specific end applications.

Another TG formation, consists of the transglutaminase enzyme with caseinate and maltodextrin. It can use this preparation to restructure foods that contain proteins in forms that are unavailable for reaction. With red meat, that allows processors to utilize the trim from steak. Adding TG allows formulators to restructure the trim into a portionable form. TG can also reduce salt levels in processed meats. It works similarly in seafood. Another type of TG, improves texture in products where there is enough protein available for the reaction to occur, such as an emulsified meat product. The preparation also works in dairy systems. And another type, which consists of transglutaminase with maltodextrin, starch and sodium phosphate, modifies the texture of seafood like block set surimi, enabling the use of lower grades of surimi while maintaining similar texture.

In yogurt, one of the biggest effects of transglutaminase is that it reduces syneresis. In frozen desserts, it changes the mouthfeel properties, especially in low-calorie or sugar-free products. Grains and cereal-based products such as pasta are potential applications. Ajinomoto is planning to release additional preparations for use in these and other food applications. The TG for use in modified fat and/ or sodium processed meat and poultry products and all non-standardized processed meat and poultry products and all types of processed seafood, natural cheese, processed cheese, cream cheese, yogurt and frozen desserts at various levels in ppm level.

Stern Enzyme

Its the new enzyme for bakery products, fruit juices and confectionery. In the modern baking laboratories,it developed enzyme components and their applications. The enzymes for the bakery trade, have created standards recognised all over the world. New products are enzymes to keep bread fresh longer and enzyme preparations to improve and rationalize the manufacturing process for biscuits, crackers and wafers.

For Baking Trials

In addition to the development of new products have been further evaluated to determine how effective they are in the production of a standard white loaf. Once the bakes were complete the products were analysed for their crumb structure, colour, texture, taste and physical parameters such as loaf height and oven spring. The new information from these trials, not just about the finished product, but also with regard to the ease of processing of the dough.This is exemplified by the xylanase story. We'll examine a wide range of baking xylanase enzymes in here. It is known that the performance of the different types of xylanase enzyme is highly dependent on the process it is used in. In addition the effect of other enzymes on the dough is greatly impacted by the type of xylanase enzyme used. The Bacillus sp. enzyme, gave a soft, workable dough which was easily machined and gave rise to a loaf with an open structure, typical of a European type bread. The Aspergillus enzyme, gave rise to a strong dough, which was easy to machine and produced a loaf with a tighter crumb and a uniform texture. The strength of the dough produced by Aspergillus enzyme suggests that this enzyme would be particularly suited for use in longer fermentation processes, and particularly for overnight fermentations where excessive dough softening can be a problem. The third enzyme, the Trichoderma sp. Xylanase, gave rise to a much softer, silkier dough which is exactly that required in no-time fermentation processes and the Chorleywood process. Oxidase enzymes for industries other than baking, and that was all trialled as baking additives. The success of the baking Glucose Oxidase, to providing the industry with standardised products for optimal activity and act in the early stages of the process, before oxygen becomes limiting. The end result is a product that has a very soft texture over an extended period preserving.

Enzyme Modified Cheese

Cheese is a major staple food of the Western World. Cheese is also an enzyme produced food. Not only does an enzyme turn the milk initially into the curds and whey creating the texture of cheese, but a whole range of enzymes turn the bland immature cheese curd into the wide range of cheese flavours that we have in the world today. With the exception of Italian cheeses, where their sharp flavours are produced by the addition of a pre-gastric esterase, the remaining types of flavours are generated by enzymes produced from the microflora present in the cheese. Moulds such as Penicillium produce the blue type cheeses such as Danish Blue (fatty acid flavour) and Roquefort (methyl ketones); and bacteria such as Lactobacillus produce protein notes in cheeses such as Cheddar. A different and more stable form of cheese is processed cheese. Processed cheese is exactly what it sounds, real fresh cheese processed in to a more stable form. Processed cheese was developed independently in Europe and the USA at the beginning of the last century. All types of cheese can be used to make processed cheese; but the different types will have a big impact on the final flavour. If very young immature cheese, which has almost zero flavour, is used then the processed cheese will also have almost zero flavour. The main advantage of using young cheese is that it is much cheaper than mature cheese. If young cheese is used, then to avoid a bland processed cheese product being made, additional flavour has to be added. This can be done in one of two ways; either a certain percentage of strongly flavoured mature cheese can be used or Enzyme Modified Cheese (EMC) can be used.

The main steps in processed cheese manufacture;

i) Blending & shredding of the cheese

ii) Emulsification

iii) Thermal processing

iv) Homogenisation (optional)

v) Cooling & packing.

EMC is made in a similar way to processed cheese. Real cheese is taken and certain enzymes added. These enzymes are the same types that turn young flavourless cheese into strongly flavoured mature cheese; but they are added at a much higher dose rate. So rather than getting a strong cheese flavour after 1 to 2 years, EMC is produced in 1 to 2 days. EMC used to be made solely by use of pig pancreatic enzymes, but the production methods used now are much more sophisticated.

The main focus is on microbial and fungal enzymes as vegetarian alternatives to pancreatic lipase. Another big advantage of these enzymes over pancreatic lipase/protease is that this animal product can only ever produce one flavour profile. It can be used in different combinations to produce a wide range of different cheese flavours. Furthermore there is always a concern about possible contamination of animal products with viruses or prions. This problem does not exist with microbial & fungal enzymes.

Efficiency from Enzymes for Fruit Juice and Bread

An enzyme has developed that can be used for fruit liquefaction and as gluten substitutes in bread baking. It consists of several plant enzymes, including pectinase, cellulase, hemicellulase, arabinase and rhamnogalacturonase that are used for processing fruit into fruit juices. That enzyme performs in the same manner as gluten, which is used for giving bread its structure. Juice processing and bread baking may seem to have little in common. But both operations use enzymes and both can benefit from employing the latest developments in enzyme technology.

That new enzyme help juice producers and bakers improve the economies and quality of their processing. Juice producers can maximize yield by using enzymes for fruit liquefaction, and bakers can use new enzymes to save on the amount of gluten they add to their bread formulas. For years, producers of fruit juice have been adding pectinases to their pressed juice to clarify it. About six years ago, they started adding macerating enzymes to fruit pomace in order to extract additional soluble solids. Macerating enzymes have included pectinase and possibly hemicellulases and cellulase. In the past few years, some producers have adapted a liquefaction process in which the pomace is almost completely liquefied. Only a small percentage of fruit matter, including the seeds (by choice), is not converted to liquid and soluble solids in this process. A complex enzyme system; contains pectinase, cellulase, hemicellulase, arabinase and rhamnogalacturonase. Amylase also may be used if the presence of starch indicates it is needed.

Instead of operating at cold temperatures, as in other processes, the liquefaction process is performed at warm temperatures, which allows the enzymes to work more effectively. Decanter/centrifuge systems remove insolubles more efficiently, and there is greater throughput in membrane filter systems. Yields can be as high as 215 to 225 gallons per ton. Although recent developments in fruit liquefaction systems offer very high yields, cold maceration processes still dominate the apple juice industry. This is mainly due to the apparent higher quality juice the process offers. It is a highly advanced enzyme that employs technology similar to the hot liquefaction process while adapting it for treatment temperatures as low degrees for as little as one to three hours of enzyme contact time.

Bread Baking

The new enzyme, to help bakers cut down on the amount of gluten they must add to their bread formulas. Gluten is added to compensate for variations in the quality (strength) of flour. Gluten, the insoluble protein in flour, forms a matrix that helps create the overall structure of bread. Its presence is crucial in bread-making. The enzymes and other functional ingredients, such as specialized flour in that enzyme, help hydrolyze pentosans and mimic gluten's functionality. Pentosans can bind many times their weight in water, which impedes the dough development in baking. Xylanases, or pentosanases, hydrolyze these high molecular weight polysaccharides. If a baker normally adds 5% gluten to a dough, he can apply 0.1% to 0.2% of the enzyme to the mix and reduce the amount of gluten to 2% to 3%.

Table: *Comparison Of Yield Using Various Apple Processes (Before Filtration) Yields From*

Process	***Peels and cores slurry (B)***	***Whole apple (B)***
Traditional	142-150	160-175
Mashing	170-180	190-210
Liquefaction	215+	225+

B = after pressing and clarification.

Yields are expressed as gallons per ton, adjusted for fibre-glucose conversion and standardized to 11.0 !degrees^ BX. Brix can range between 10-17 !degrees^; 11- 11.5 !degrees^ BX is considered standard single strength.

Bibliography

Ashraf, Syed Mahboob: *Handbook of Fruit and Vegetable Products*, Agrobios, Delhi, 2008.

Avanish K. Tiwari: *Food Security and Global Economy*, Pentagon Press, Delhi, 2009.

Bajpai, Amita: *Recent Trends in Animal Nutrition and Feed Technology for Livestock, Pets and Laboratory Animals*, International, 2003.

Barbeau, G.: *Tropical Fruits in Nicaragua*, Managua, Nicaragua Ministerio de Desarrollo Agropecuario, Agraria, 1990.

Bushnell, R.B.: *Dry Cow Feeding and Management*, A Western Regional Extension Publication, 1979.

Chakraborty, Sudip: *Food Security and Child Labour : The Case of a Hazardous Occupation*, Deep and Deep, Delhi, 2011.

Cristobal Noe Aguilar: *Food Science and Food Biotechnology in Developing Countries*, Asiatech Pub, 2008.

Daphne C. Elliott: *Biochemistry and Molecular Biology*, Oxford University Press, Delhi, 2005.

Eckel, Peter J.: *College & University Foodservice Management Standards*, Westport, AVI Pub. Company, 1985.

Featherly H. I.: *Taxonomic Terminology of the Higher Plants*, USA, Iowa State College Press, 1954.

Gautam Arvind : *A Concise Course on Hotel and Hospitality Management*, Axis Pub, Delhi, 2010.

Gautam, U.S. and Ram Chand: *Decision-Making of Dairy Beneficiaries: Role of Aspiration, Motivation and Knowledge*, Om Pub, Delhi, 2010.

Hacker, J.B.: *Nutritional Limits to Animal Production from Pasture,* Farnham Royal: CAB, 1981.

Jha, S N: *Dairy and Food Processing Plant Maintenance : Theory and Practice,* International Book Distributing, Delhi, 2006.

Khurana A D : *Handbook of Catering, Hospitality and Tourism,* Mohit Pub, Delhi, 2008.

Krimsky, Sheldon: *Biotechnics and Society: The Rise of Industrial Genetics*, New York: Praeger Publishers, 1991.

Law, Barry A.: *Microbiology and Biochemistry of Cheese and Fermented Milk*, London: Blackie Academic & Professional, 1997.

Lois D. McBean.: *Handbook of Dairy Foods and Nutrition*, Boca Raton, Fla.: CRC Press, 1999.

Michael, J. Lewis,: *SeparationProcesses in the Food and Biotechnology Industries*, Cambridge, U.K.: Woodhead, 1996.

Mowery, D. C. : *Technology and Wealth of Nations*, Stanford: Stanford University Press, 1992.

Naren Kumar Dutta: *Fundamentals of Biochemistry : A Practical Approach*, Kanishka, Delhi, 2005.

Paul B. : *Food Biotechnology in Ethical Perspective*, Aspen, CO: Aspen Publishers, 1997.

Rao, P. Venkateshwara: *Dairy Farm Business Management*, Biotech Books, Delhi, 2008.

Robert C.: *Cases in Hospitality Marketing and Management*, New York, John Wiley, 1997.

Spreer, Edgar: *Milk and Dairy Product Technology*, Translated by Axel Mixa. New York: M. Dekker, 1998.

Thompson, Paul B.: *Food Biotechnology in Ethical Perspective*, Aspen, CO: Aspen Publishers, 1997.

Upadhyay, K.G. and Vyas, S.H.: *Composition of Camel's Milk*, Gujarat Agric. University, 1982.

Vikram V. Mistry: *Cheese and Fermented Milk Foods*, Great Falls, Va.: Kosikowski, 1997.

William H.: *The Chef's Guide to Practical Restaurant Cookery*, New York, Van Nostrand Reinhold, 1988.

Index

❑❑❑